苏州大学百年建筑

Historical Buildings at Soochow University

吴永发 戴叶子 钱晓冬 编著

苏州大学出版社
Soochow University Press

图书在版编目(CIP)数据

苏州大学百年建筑 / 吴永发，戴叶子，钱晓冬编著 .
—苏州：苏州大学出版社：2014.6
ISBN 987-7-5672-0910-7

Ⅰ . ①苏… Ⅱ . ①吴… ②戴… ③钱… Ⅲ . ①高等学校 - 教育建筑 - 苏州市 - 图集 Ⅳ . ① TU244.3-64
中国版本图书馆 CIP 数据核字（2014）第 105428 号

本书为苏州大学文化发展战略实施专项资助项目

书　　名：苏州大学百年建筑 (Historical Buildings at Soochow University)
编　　著：吴永发　戴叶子　钱晓冬
责任编辑：史创新　沈　琴
外文编辑：滕育梅　David Kindred
出版发行：苏州大学出版社 (Soochow University Press)
社　　址：苏州市十梓街 1 号　邮编：215006
网　　址：www.sudapress.com
印　　刷：苏州工业园区美柯乐制版印务有限责任公司（电话号码 0512-67606001）
邮购热线：0512-67480030
销售热线：0512-65225020
开　　本：889mm×1194mm　1/12　印张：6.5　字数：75 千
版　　次：2014 年 6 月第 1 版
印　　次：2014 年 6 月第 1 次印刷
书　　号：978-7-5672-0910-7
定　　价：88.00 元

由购本社图书发现印装错误，请与本社联系调换。服务热线：0512-65225020

前言

苏州是一座具有2500多年建城历史的文化名城，这里曾经是春秋吴国的都城，素有“人间天堂”的美称。一个多世纪前，中国著名的现代高校东吴大学在这里创建，开中国西式教育之先河，成为苏州大学的最早前身。

从早年的中国名校东吴大学到省内外享有盛誉的江苏师范学院，又从江苏师范学院到国家“211工程”重点建设的苏州大学，百余年来，苏州大学历尽沧桑，迭经变革，栉风沐雨，创建辉煌。

悠久的发展历史，在苏州大学校园中积累了为数众多的具有中国古典特色、民国特色及欧式特色的校园建筑，形成了苏州大学鲜明的校园风貌。2013年5月，苏州大学天赐庄校区东吴大学旧址作为近现代史迹及代表性建筑，入选第七批全国重点文保单位。以此为契机，苏州大学建筑系重新整理、测绘了校园内的优秀近代建筑，同时辅以与这些建筑相关的部分历史照片，形成了现在我们看到的《苏州大学百年建筑》。这些精致的图纸和珍贵的图片展示着学校优良的办学传统，彰显着学校深厚的文化底蕴，勾画了“天堂学府”苏州大学百年发展的历史轨迹。

愿苏州大学这棵百年老树更加青春焕发，根深叶茂，繁花似锦，硕果满枝！

PROLOGUE

Suzhou, once the capital of the Wu Kingdom in the Spring and Autumn Period, is a city with profound culture and a history of more than 2,500 years. Its reputation as "Paradise on Earth" in China is well deserved. More than a century ago, Dongwu University was founded as an institute of renowned modern higher education in China, becoming the first institute with western style education in China. It was from these roots that Soochow University evolved.

Just as the course of a mighty river is carved out and given form by all that has moved through it, Soochow University is a product of many shaping events and continues to change and reform with the passing of every day.

Over the past century, the shapers of this institute were and are the people who hold the school near and dear to their hearts. The educators, the caring administrators and the students and graduates all have guided the school through its various iterations from Dongwu University, the nationally renowned Jiangsu Normal College and finally to its 21st Century form as Soochow University, one of the top 100 key universities in China.

With a long history, Tian Ci Zhuang campus of Soochow University has accumulated a large number of buildings. These buildings have classical characteristics, Mingguo style, and European features, forming a distinctive campus of Soochow University. In May, 2013, campus (the former site of Dongwu University) was selected as being representative of modern historical sites, was given state-level protection as part of the 7th batch of Cultural Heritage Sites.

In recognition of this great preservation achievement, the Department of Architecture at Soochow University undertook a survey of the buildings at Soochow University and is proud to present "Historical Buildings at Soochow University", an album of exquisite drawings and precious images, depicting the rich cultural heritage and traditions of Soochow University over the past 100 years.

It is our fervent wish that Soochow University will continue to grow and prosper. We pass on our love for our school to the next generation and trust its preservation in your hands.

苏州大学沿革　HISTORY OF SOOCHOW UNIVERSITY

苏州大学的最早前身为创建于1900年的东吴大学；1952年全国院系调整时，东吴大学、苏南文化教育学院、江南大学合并组建苏南师范学院，同年更名为江苏师范学院，在原东吴大学校址办学；1982年江苏师范学院改为苏州大学。苏州蚕桑专科学校、苏州丝绸工学院、苏州医学院先后于1995年、1997年和2000年并入苏州大学。

Dongwu University, founded in 1900, is the most notable predecessor of modern-day Soochow University.

In 1952, Dongwu University, Southern Jiangsu College of Culture and Education, and Jiangnan University were merged to create the Southern Jiangsu Normal College. Later in that same year, it was renamed as Jiangsu Normal College and it was situated on the old campus of Dongwu University.

Thirty years later, in 1982, Jiangsu Normal College was reorganized and renamed Soochow University.

In the years of 1995, 1997 and 2000, Suzhou College of Sericulture, Suzhou Institute of Silk Textile Technology and Suzhou Medical College were brought into the Soochow University family.

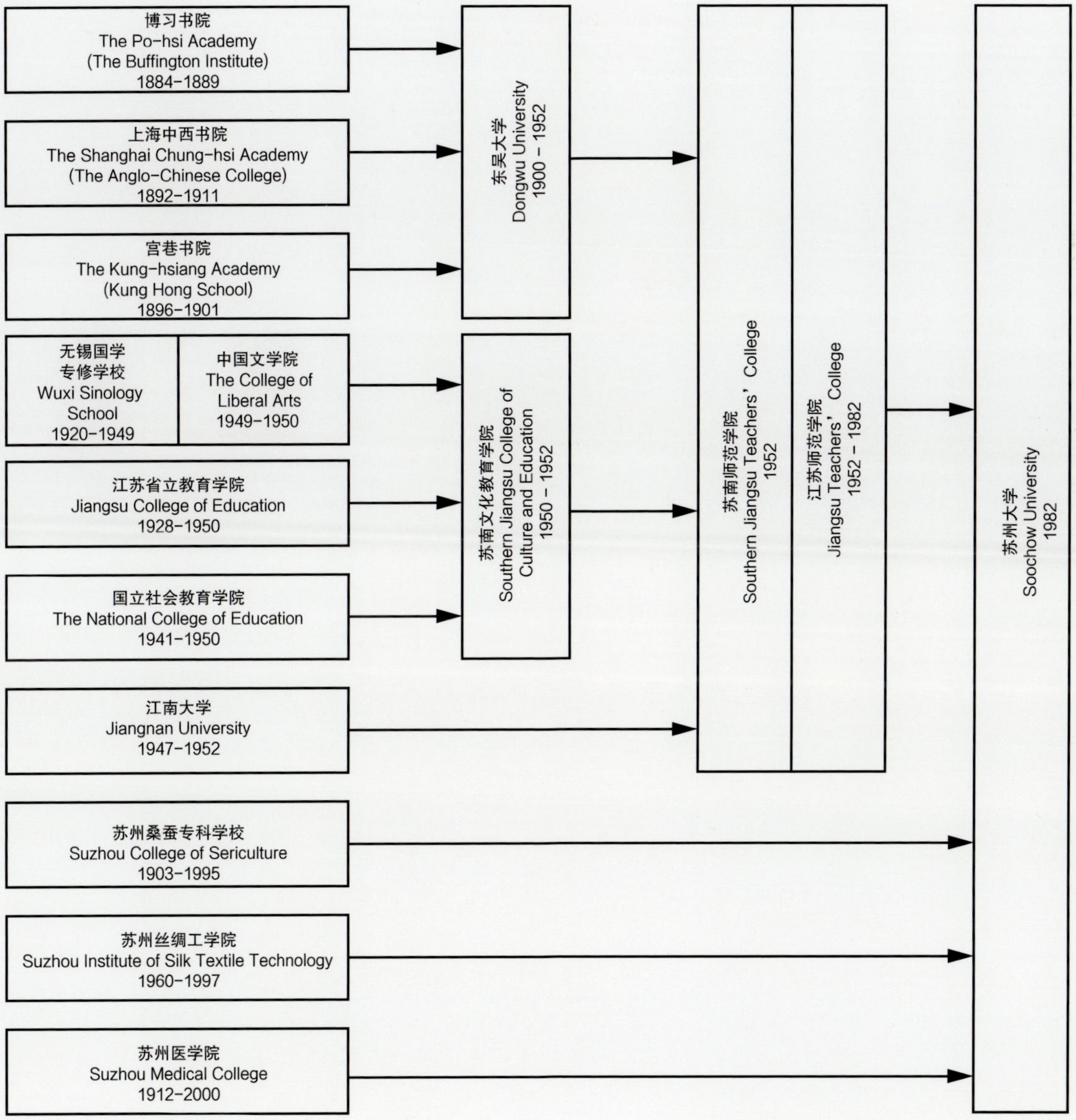

苏州大学历史建筑分布图
PLAN OF HISTORICAL BUILDINGS

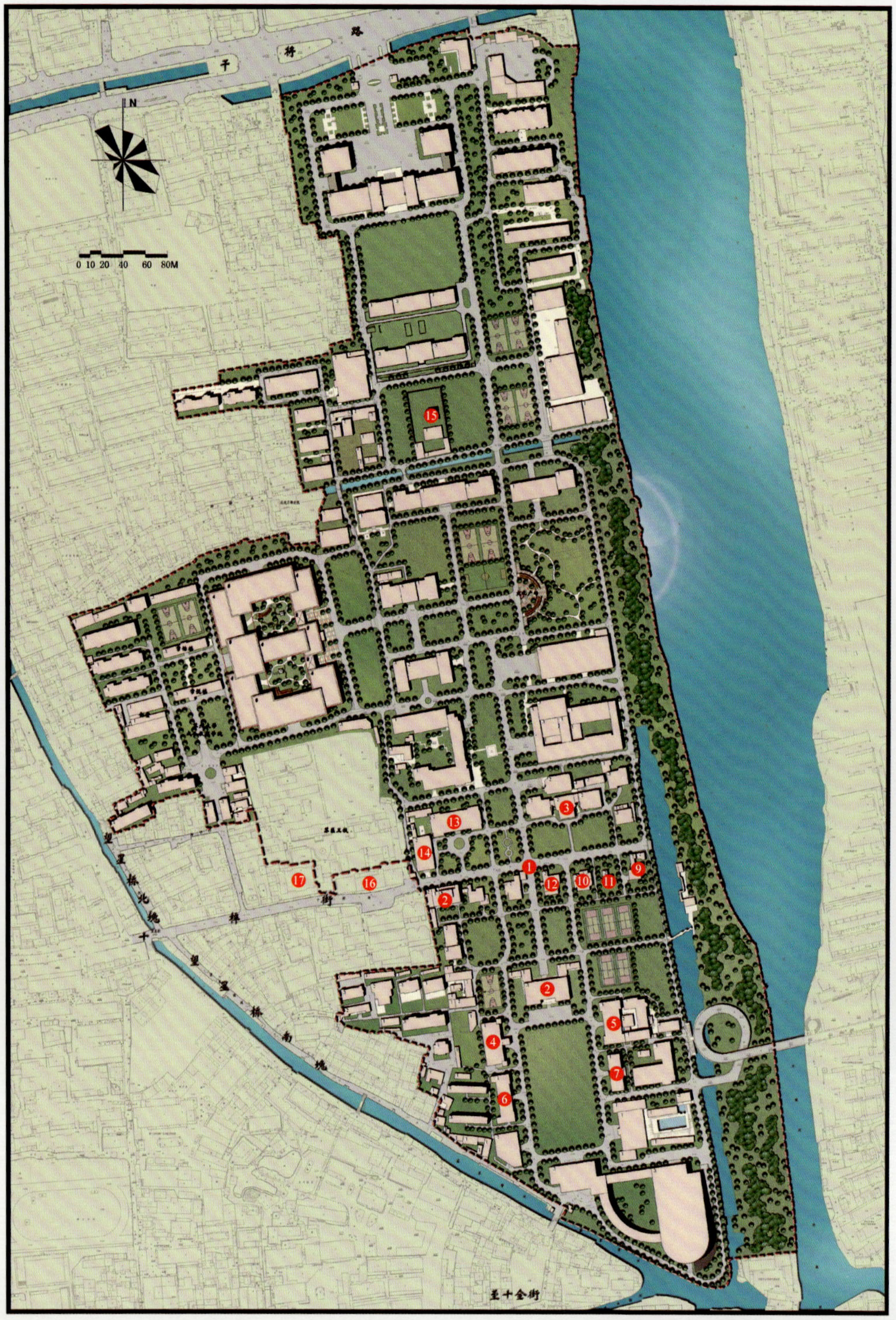

苏州大学天赐庄校区总平面图
General Plan of Tian Ci Zhuang Campus

1.东吴大学校门(The Old Gate of Dongwu University)
2.钟楼(The Clock Tower Building)
3.红楼(The Red Mansion)
4.精正楼(Jingzheng Building)
5.维正楼(Weizheng Building)
6.维格堂(Weige Hall)
7.子实堂(Zishi Hall)
8.博物馆(The Soochow University Museum)
9.春晖楼(Chunhui [Spring] Building)
10.夏润楼(Xiarun [Summer] Building)
11.秋韵楼(Qiuyun [Autumn] Building)
12.冬瑞楼(Dongrui [Winter] Building)
13.崇远楼(Chongyuan Building)
14.敬贤堂(Jingxian Hall)
15.文星阁(Wenxing Tower)
16.博爱楼 (Bo' ai Building)
17.圣约翰堂(St. John's Church)
18.可园(Ke Garden)(位于天赐庄校区西南2公里的苏州大学南校区)

目录 CONTENTS

东吴大学校门　THE OLD GATE OF DONGWU UNIVERSITY

建筑年代:始建于20世纪初,重建于1948年

历史沿革:东吴大学校门坐南朝北,校门以南便是原东吴大学办学校址,也是本部建筑精华所在。

原校门为巴洛克风格,"东吴大学"校名系翁同龢题。大门南面镌有中英文校训"UNTO A FULL GROWN MAN"(新中国成立后改为"全心全意为人民服务")。1935年加刻校训"养天地正气,法古今完人"。目前大门为1928级校友于1948年5月捐建。

东吴大学校门是很多造访苏大的来宾都愿意为之驻足的标志性建筑,这不仅因为它有着洋气的三个圆拱门设计,更因为它串联起了人们对东吴大学这所在中国高校办学史上有着特殊地位的大学的记忆和联想。东吴大学校门的特别之处在于能引导来访者的视线落在苏大最美丽的所在,穿过圆门,钟楼、精正楼、维格堂、子实堂等风格各异的建筑渐次映入眼帘,犹如一个小小的建筑博览会。

Years of construction: originally built in the early 1900's, rebuilt in 1948

History: The Gate is north-facing. The original site of Dongwu University is to the south of the gate where many of the most architecturally profound buildings are still located today.

The original name, "Dongwu University", was bestowed upon the school by Mr. Weng Tonghe, a Confucian scholar and imperial tutor during the Qing Dynasty.

The old gate is of Baroque style. Its original name, "Unto a Full-Grown Man", was originally carved on the south facet of the Gate. It was later changed to "Serve the People Wholeheartedly" after the founding of the People's Republic of China.

In 1935, a Chinese version of the school motto, "養天地正氣,法古今完人", was carved to reflect the English motto "Unto a Full-Grown Man"。With generous donations by the alumni of the class of 1928, the current Gate was rebuilt in 1948.

The old gate of Dongwu University is symbolic of the architecture of Soochow University and is probably the most photographed site on campus, serving as a backdrop for photographs of the many visitors from all over the world.

It has three modern arches and it instantly stirs the memories of anyone who was associated with the school, which played such a pivotal role in the establishment of the Chinese post-secondary education system.

Through the arches of the gate, a visitor is guided to the most beautiful scene on campus. The Clock Tower, The Jingzheng Building, Weige Hall, and Zishi Hall all strike a note of awe for any visitor. It is like stepping into an architectural exhibition of unique buildings found nowhere else in China.

东吴大学校门历史照片
Vintage Image

东吴大学校门北立面石狮及题字
North Stone Lion and Inscription

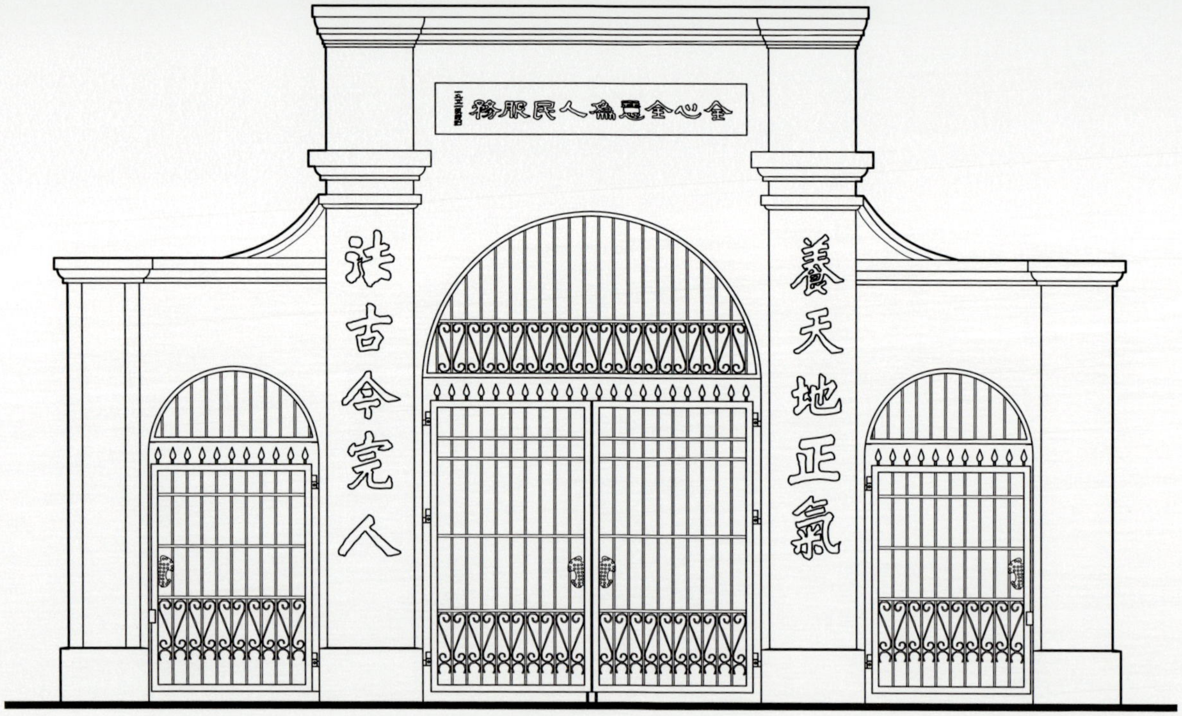

东吴大学校门南立面
South Elevation

东吴大学校门南面全景
South View

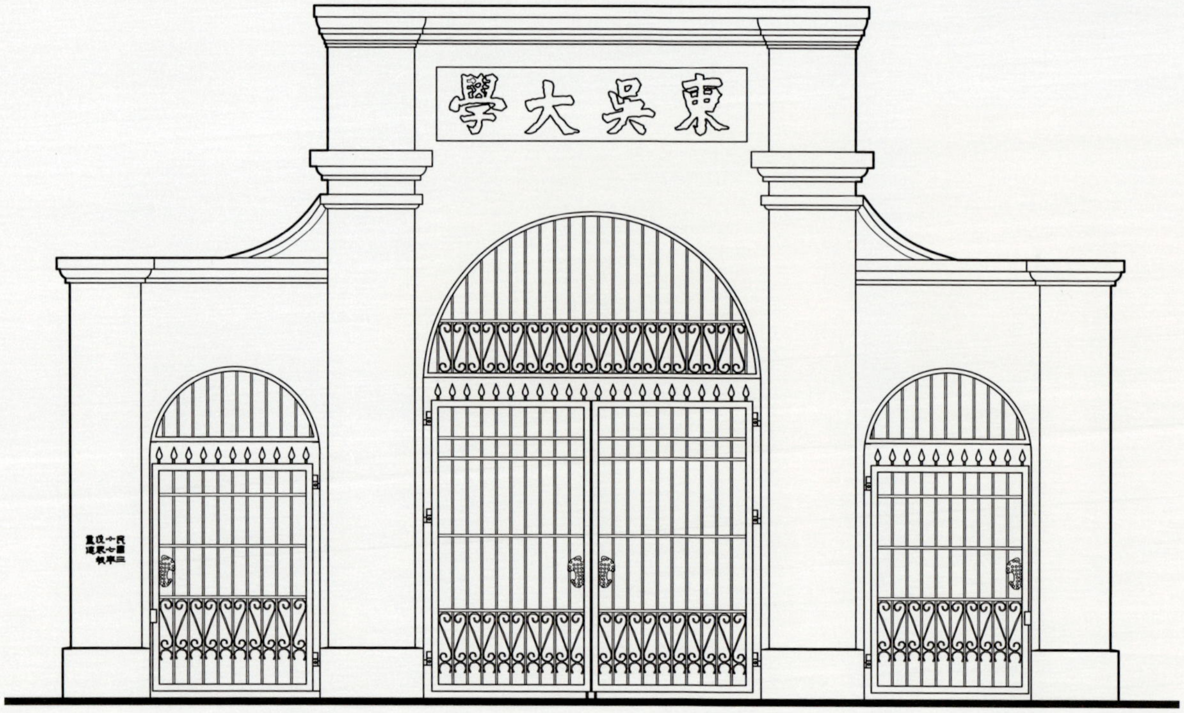

东吴大学校门北立面
North Elevation

东吴大学校门北面全景
North View

钟楼 THE CLOCK TOWER BUILDING

建筑年代：始建于1901—1903年

原名：林堂

原功能：图书馆、礼堂、授课室及试验室，新中国成立后为外语楼

现名：钟楼

现功能：行政办公

历史沿革：目前校内最古老的教学主楼。原名林堂，为纪念该校奠基者之一林乐知得名。因建筑上层的钟塔位于校区的中轴线上，顶部置有报时大钟，所以又被称为钟楼。该楼落成后，成为当时苏州地区规模最大的西式建筑。作为教会学校，东吴大学早期设有神学课程，师生的祈祷、布道活动多于此处进行。

钟楼以红砖勾勒框架和窗楣，青砖填充墙面，而柱式、线脚和花饰则为石制。玫瑰窗、花式窗棂及钟塔等建筑元素凸显出早期东吴大学教会学校的身份，也令其在一般的西式建筑中显得不同凡响。高高耸立的钟塔在外廊和两边壁柱的衬托下显得优雅挺拔，青红砖墙与石材条带组合而成的色彩基调沉着而明快。

钟楼在建筑形式上隽秀雅致，比例优美精当。无论是在当时还是历经百十年后的今天，它都是整个学校的标志性建筑。

Years of construction: 1901-1903
Original name: Lin Tang(Allen Hall)
Original usage: library, auditorium, classroom, lab, offices of the Foreign Language Department
Current name: The Clock Tower Building
Current usage: administration offices
History: The Clock Tower Building is the oldest building on campus. It was originally named Lin Tang (Allen Hall), to commemorate Lin Yuezhi (Dr. Young J Allen) who was one of the founders of the school.
The clock tower atop the building is in line with the axis of the campus and has a big chiming clock built in which gives the Lin Tang building its current English name, "The Clock Tower Building".
When it was built, it was the largest western style building in Suzhou.
As a Methodist School, Dongwu University offered theology courses. Some activities such as church services for teachers and students, as well as sermons were often carried out here.
The Clock Tower Building is of grey brick with stone columns and ornaments and the frame and window lintels are outlined with red brickwork.
The rose windows, the floral window lintels, and the clock tower are the main architectural elements of the building, and clearly identify its origin as a church school.
The clock tower itself stands straight and elegantly against the sky.
With pilasters and exterior aisles, the red and grey brick walls blend harmoniously with the stone rims, giving the building a delicate and crisp tone.
The building is well proportioned in architectural design, handsome and elegant.
It has been and always will be the symbolic architecture of the university.

钟楼南面全景
South View

钟楼北立面
North Elevation

钟楼南立面
South Elevation

钟楼西立面
West Elevation

钟楼南立面玫瑰窗
Rose Window on South Facade

钟楼历史照片
Vintage Image

钟楼北面全景
North View

钟楼北面拱券柱头
Capital of North Porch

钟楼南面玫瑰窗
South Rose Window

钟楼北面柱廊
North Arcade

钟楼北面柱廊柱头
Capital of North Arcade

钟楼北面门斗
North Porch

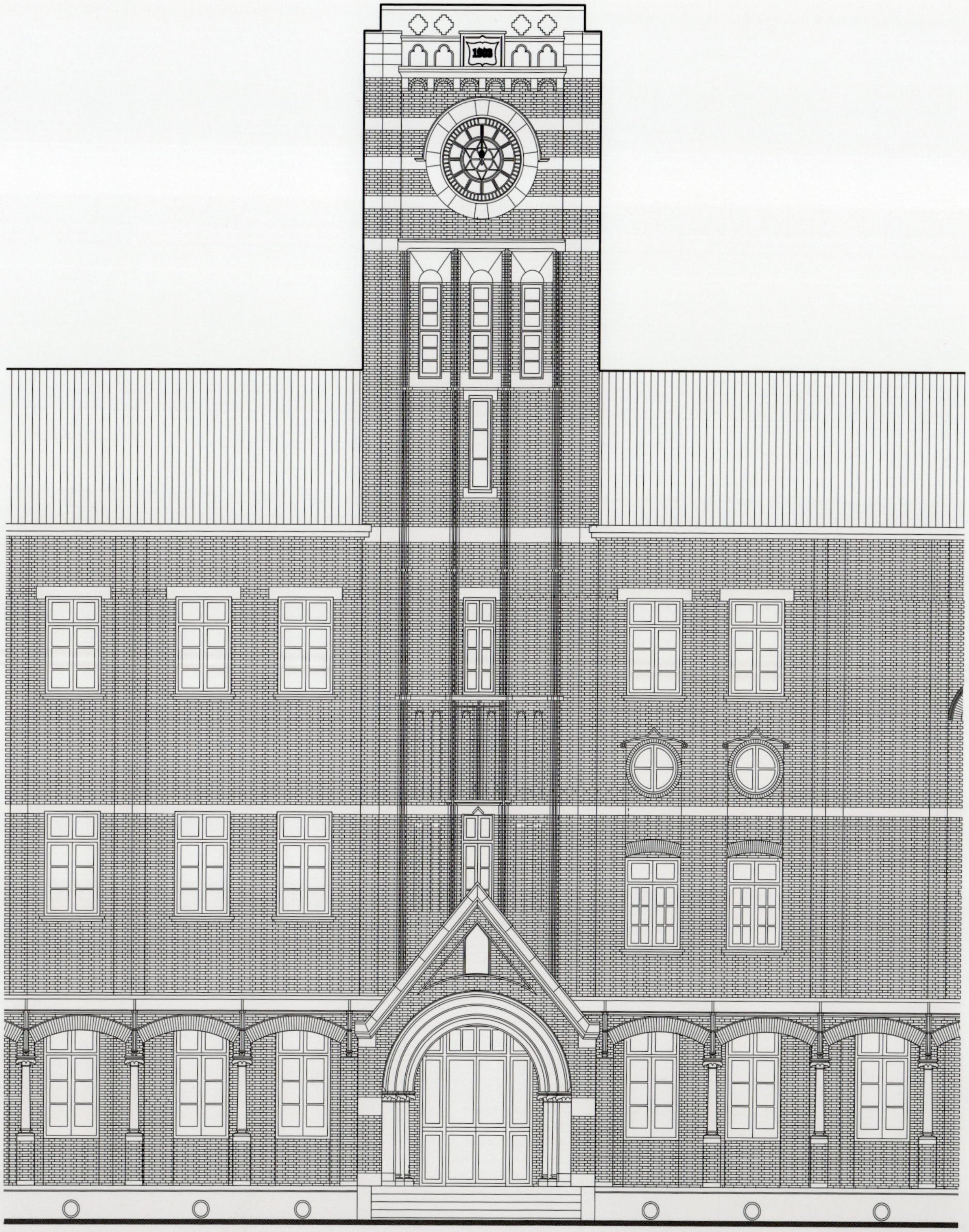

钟楼北立面中部
Midsection of North Elevation

钟楼南立面中部
Midsection of South Elevation

钟楼南面中部
Midsection of South Facade

红楼　THE RED MANSION

建筑年代：始建于1902年

原名：景海女子师范学校

原功能：景海女子师范学校办公、住宿、教堂

现名：苏州大学红楼会议中心

现功能：会议、接待

历史沿革：苏州景海女子师范学校前身是于1902年(光绪二十八年)在苏州天赐庄创办的景海女塾，原为纪念美国基督教监理公会女传教士海淑德(Laura Askew Haygood)而建。1917年(民国6年)改为景海女子师范学校。新中国成立后，学校经分拆合并后不复存在。2004年12月23日，景海女师旧址被列为苏州市第五批公布的市级文物保护单位。

Year of construction: 1902

Original name: Jinghai Female Normal School

Original usage: the school office, accommodation, Chapel

Current name: The Red Mansion Conference Centre of Soochow University

Current usage: conferences, reception

History: The "Red Mansion" was originally built to house the Jinghai Female Private School which was founded by Suzhou Tian Ci Zhuang in 1902. It commemorates Laura Askew Haygood, an American commissioner sent by the Women's Missionary Society of the Methodist Church. Later in 1917, it was renamed "Jinghai Female Normal School". The school was split and re-united several times after the founding of the People's Republic of China, but now, it no longer exists. The old site of Jinghai Female Normal School was listed in the fifth batch of Municipal Heritage Sites in December 2004.

原景海女子师范学校总平面图
Original Plan of Former Jinghai Female Normal School

红楼南面全景
South View

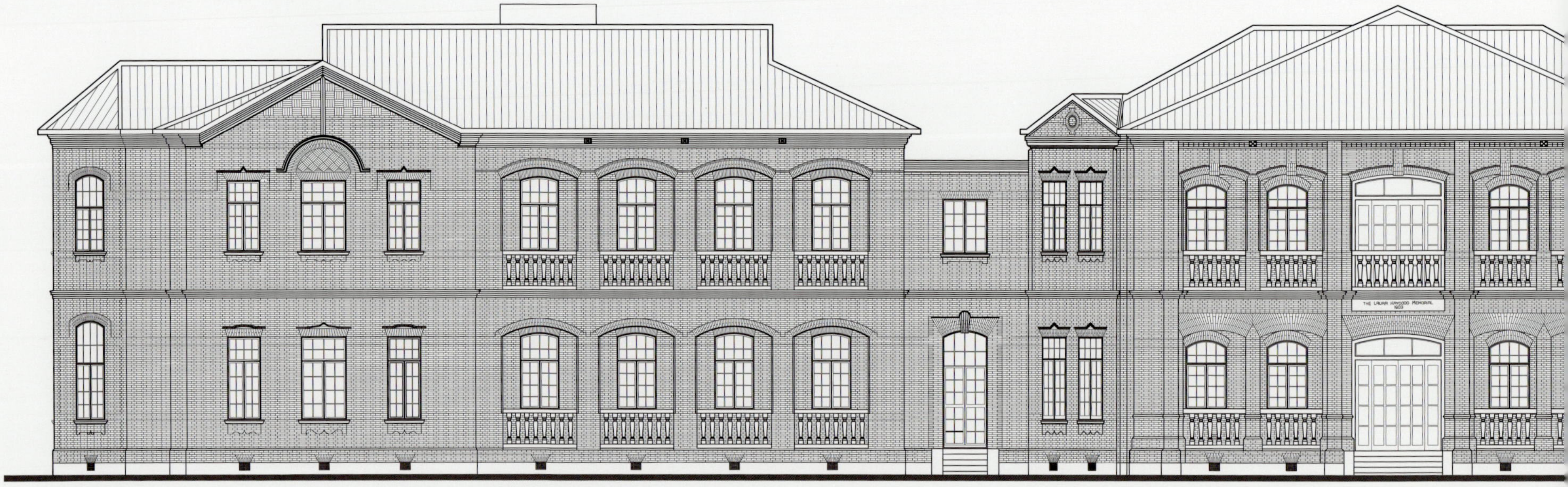

该区域连接加建学术报告厅
Connecting Passageway to Conference Hall (addition)

红楼南面门廊
South Porch

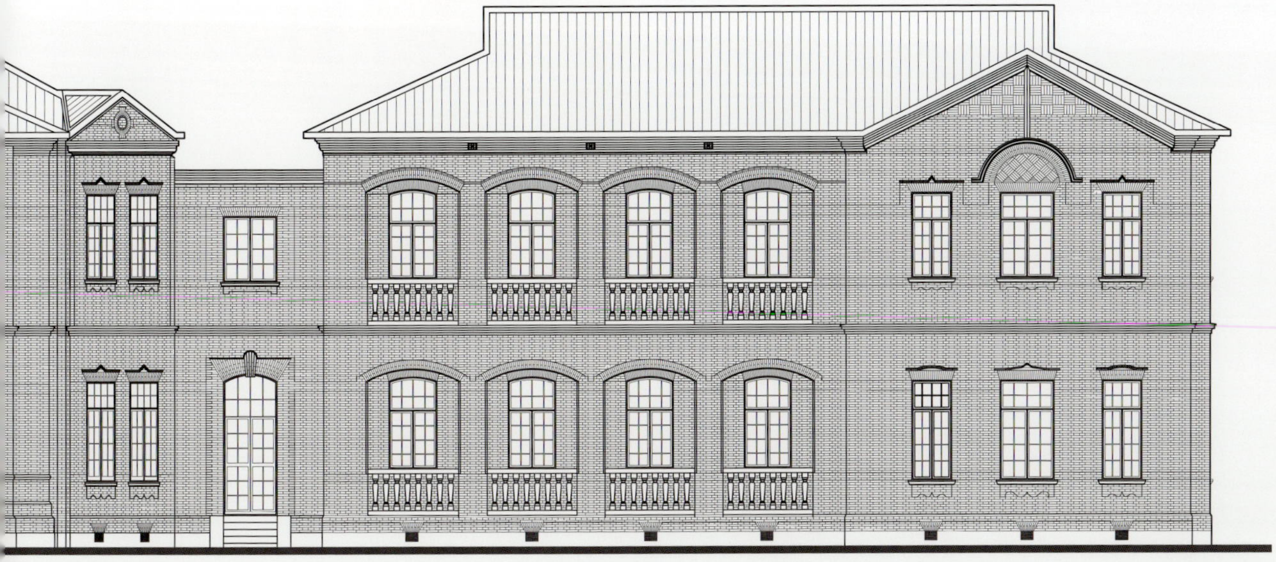

红楼南立面
South Elevation

红楼南面局部
Detail of South Facade

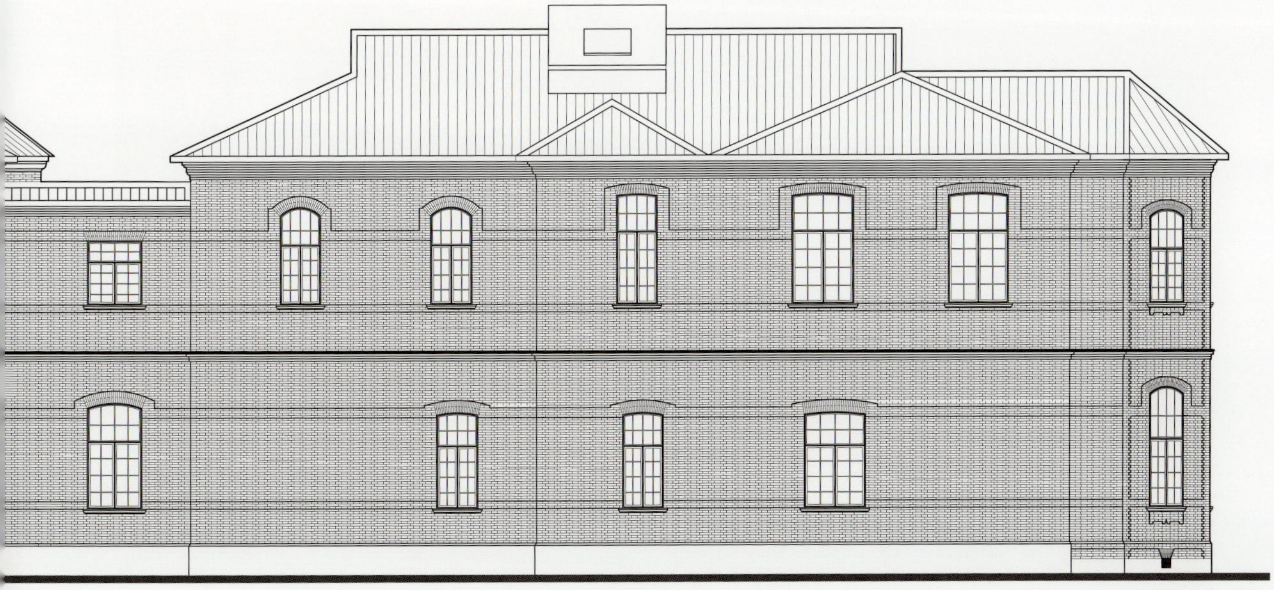

红楼北立面
North Elevation

红楼南面局部
Detail of South Facade

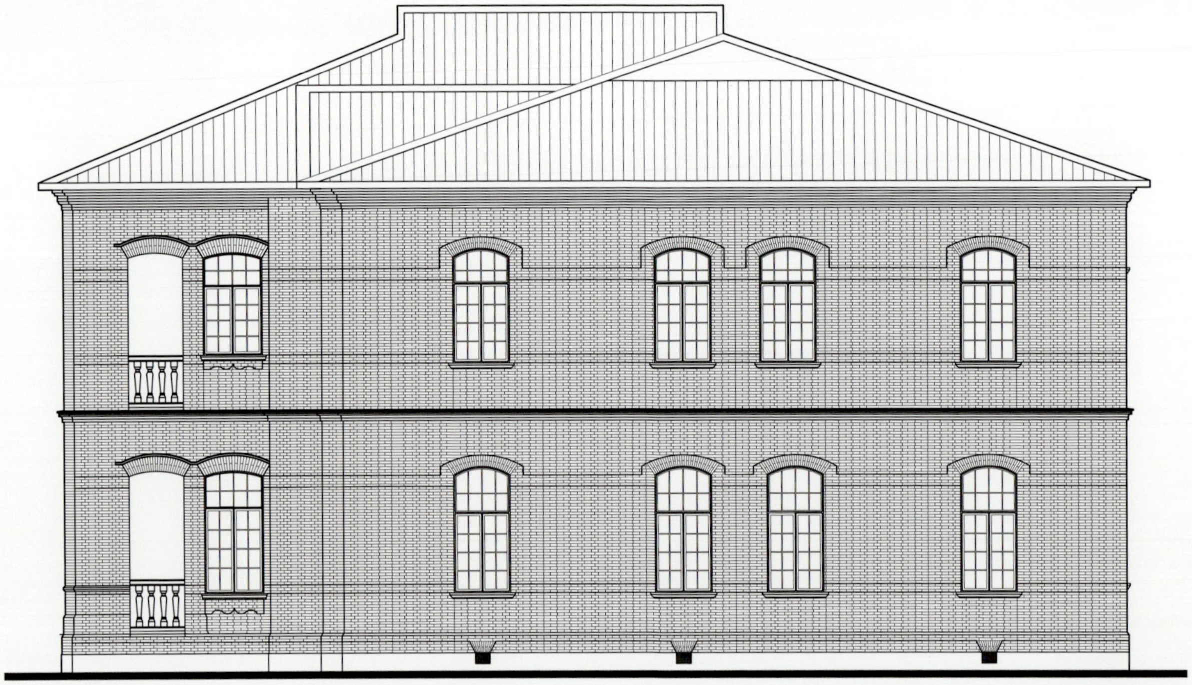

红楼东立面
East Elevation

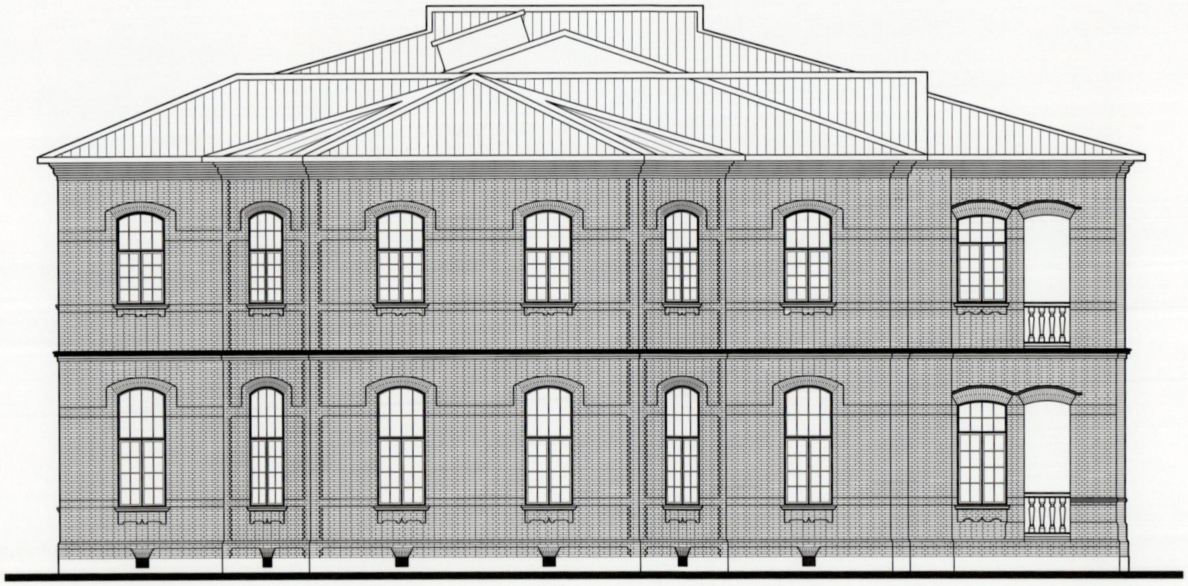

红楼西立面
West Elevation

红楼西面局部
Detail of West Facade

红楼东面局部
Detail of East Facade

红楼南面门廊局部
Detail of South Porch

红楼南面中部
Midsection of South Facade

红楼南立面中部
Midsection of South Elevation

精正楼 JINGZHENG BUILDING

建筑年代：1908—1912年

原名：孙堂

原功能：最初内设藏书馆、中西文授课室、职员办公室、化学实验室及大会堂等用房

现名：精正楼

现功能：数学科学学院

历史沿革：由弗吉尼亚州林企堡考特街区教堂捐建。1911年，该馆正在建设中，原东吴大学第一任校长孙乐文先生(Dr. David L. Anderson)去世。为纪念他，该馆建成后被命名为"孙堂"。

精正楼的平面布局与立面构图均基本对称，建筑风格细腻，施工工艺精良，风格上以英国哥特复兴式为主。入口门廊为哥特式尖券造型，高两层，门洞上部缀有精美的石雕花饰，是近代西式风格建筑中难得的精品。夏日的精正楼立面爬满了藤蔓，与美国东部常春藤大学校舍的面貌十分相似，尤其是在斜阳夕照下，整座建筑备显高贵、神秘与幽静。

Years of construction: 1908-1912

Original name: Sun Tang (Anderson Hall)

Original usage: library, Chinese and foreign language classrooms, chemistry lab, auditorium

Current name: Jingzheng Building

Current usage: School of Mathematical Science

History: The Jingzheng Building was donated by the Court Street Church of Lynchburg, Virginia in the United States. In 1911, while the building was still under construction, David L. Anderson, first president of Dongwu University died, so the donors named it Anderson Hall as a memorial. The Chinese name of the building, Sun Tang was adopted from Mr. Anderson's Chinese name, Lewen Sun.

The plan and elevation of the Jingzheng Building are symmetrical. The sophisticated construction and craftsmanship of the building is in British Gothic Revival style with artistic stone carvings at the entrance leading to a double-storey hall entrance featuring a Gothic arch, a true masterpiece of modern western architecture.

In summer, the main facade is covered with vines, reminiscent of the campuses of Ivy League schools in the United States.

Sunset is a particularly beautiful time to visit when the Jingzheng Building exudes elegance, quiet and mystery.

精正楼东面全景
East View

精正楼历史照片
Vintage Image

精正楼东门廊
East Porch

精正楼南面底层窗户
South Window (First Floor)

精正楼南面二层窗户
South Window (Second Floor)

精正楼南立面
South Elevation

精正楼东立面
East Elevation

精正楼历史照片
Vintage Image

精正楼东门廊柱头
Capital of East Porch

精正楼东面门斗
East Porch

精正楼西立面
West Elevation

精正楼北塔楼局部
Detail of North Tower

精正楼西面老虎窗
West Dormer Window

精正楼西面局部
Detail of West Facade

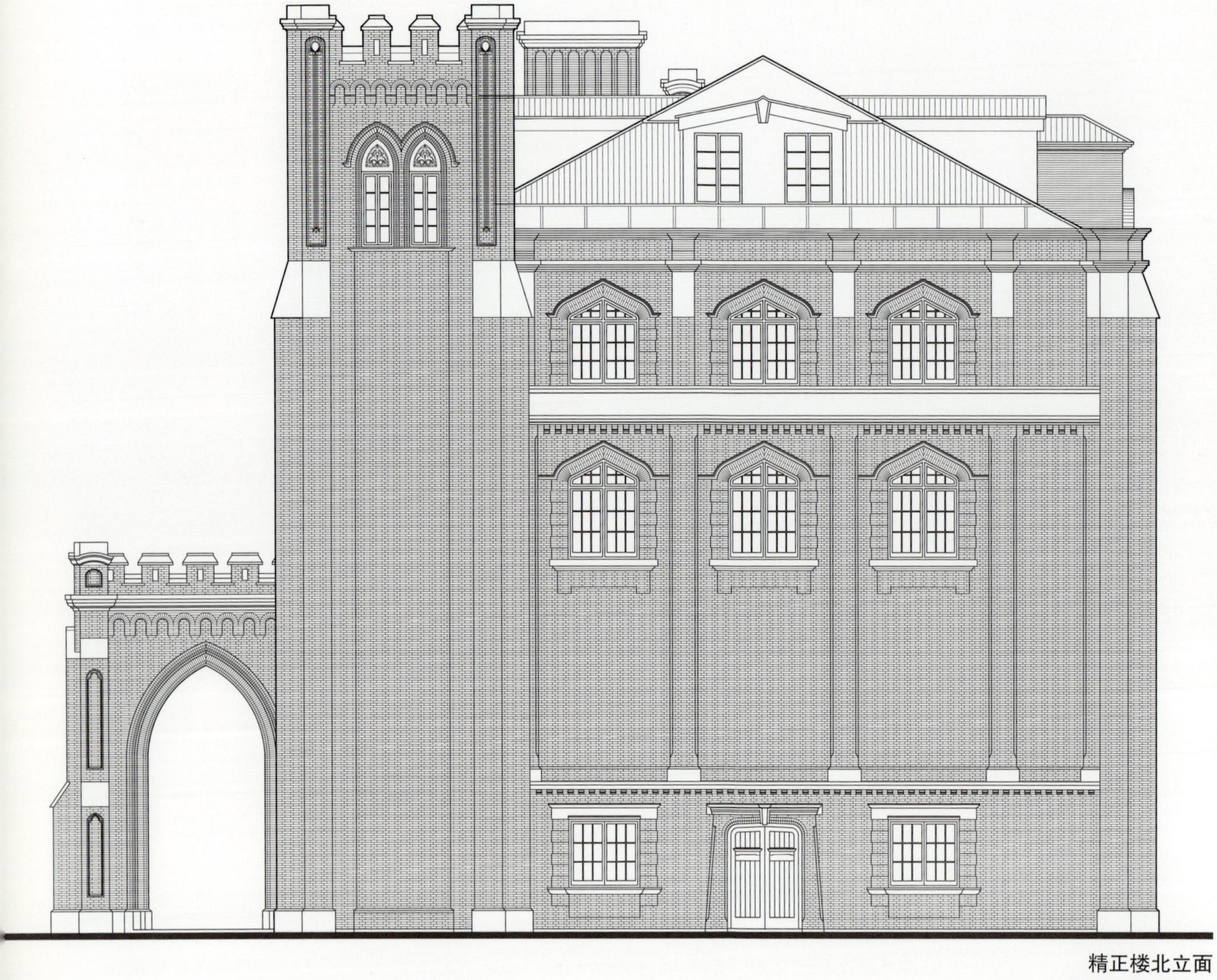

精正楼北立面
North Elevation

精正楼东门廊北面
East Porch(North Facade)

精正楼北面四层窗套
North Window(Forth Floor)

精正楼北面底层
North Facade(First Floor)

精正楼东面中部
Midsection of East Facade

精正楼东立面中部
Midsection of East Elevation

精正楼西立面中部
Midsection of West Elevation

精正楼西面中部
Midsection of West Facade

维正楼　WEIZHENG BUILDING

建筑年代：1922—1923年

原名：葛堂

原功能：理科大楼，生物、化学、数学和物理学科

现名：维正楼

现功能：苏州大学护理学院

历史沿革：子实堂北面是维正楼，原名葛堂。1920年，美国阿肯色州小石城第一监理会教堂向东吴大学捐赠65000美元，用于建筑一幢理科大楼。这是东吴大学自开办以来接受的最大的一笔捐款。用此捐赠所建的理科大楼被命名为"葛堂"，以纪念东吴大学第二任校长葛赉恩的父亲。

葛堂外观方正朴实，以竖向线条作三段式处理，以哥特式扶壁和尖拱门洞装饰突出强调正中的入口大门。建成时顶部有哥特式的小尖塔，日久损坏。葛堂同样以红砖砌筑，造型简洁。

Years of construction: 1922-1923

Original name: Ge Tang (Cline Hall)

Original usage: the building of Science, including the subject of biology, chemistry, mathematics and physics

Current name: Weizheng Building

Current usage: School of Nursing

History: To the east of Zishi Hall stands the Weizheng Building which was originally named Ge Tang. In 1920, The First Methodist Church of Rock City, Arkansas, donated $65,000 USD to Dongwu University for the construction of the first building dedicated to Science. At that time, it was the largest donation that Dongwu University had received since the opening of the school.

The building was originally named "Ge Tang" (Cline Hall) in memory of the father of J. W. Cline [Chinese name, 葛(Ge)赉(Lai)恩(En)], the second President of Dongwu University.

The facade of the building is regular and balanced, utilizing vertical lines to visually divide the building into three parts. The central entrance, decorated by a pointed arch, is emphasized by Gothic flying buttresses. The Gothic tower atop the building has been weathered. Built with red brick, the Weizheng Building neatly blends with its surroundings.

维正楼西面全景
West View

维正楼西南侧局部
Detail of South-west Facade

维正楼西立面
West Elevation

维正楼西面局部及三层老虎窗
Detail of West Facade and Dormer Window(Third Floor)

维正楼西面中部局部
Midsection of West Facade

维正楼西立面门廊局部
Detail of West Porch

维正楼西面入口
West Entrance

维格堂　WEIGE HALL

建筑年代：1931—1932年

原名：维格堂

原功能：学生宿舍

现名：维格堂

现功能：苏州大学数学科学学院

历史沿革：精正楼南边是维格堂，为纪念李维格而命名。李维格，江苏吴县(今江苏苏州)人。我国近代钢铁专家，汉阳铁厂的主要开拓者。李维格非常重视技术人才的培养，并且深深体会到，必须要有自己的技术人才，才能发展本国的产业。正是由于这种思想，他积极办学，临终前数日，他把所置产业的三分之一捐赠给东吴大学，资助贫寒学生，发展教育事业。维格堂即是李维格捐资所建。当时主要是用作男生宿舍，这样的用途一直延续到上世纪的80年代。蒋纬国两次来东吴大学求学时，都在此住宿过。

今日，经过几十年的风风雨雨，古朴而厚实的维格堂仍屹立在苏大美丽的校园内，完好无损。

Years of construction: 1931-1932

Original name: Weige Tang

Original usage: student dormitory

Current name: Weige Hall

Current usage: School of Mathematical Science

History: To the south of the Jingzheng Building stands Weige Hall, named in memory of Weige Li.

Li was born in Wu Town(the Present Suzhou City), Jiangsu Province. He was an expert in the steel industry of Modern China, and a founder of Hangyang Steel Factory. Li paid a lot attention to the professional training of the technical talents. He fully understood that China needed a wealth of home-grown talents in order to develop its own industry. With this idea in his mind, Li took an active role in China's education.

In his will, Li donated fully one third of his wealth to Dongwu University in order to help those students who could not afford the tuition.

Weige Hall was funded by a portion of Li's donation. It was originally built and used as the boy's dormitory until the 1980's. Weikuo Chiang lived at Weige Hall during his study at Dongwu University.

Although Weige Hall has endured several decades of weathering, it is still preserved in good condition and stands on the beautiful campus in a simple and modest way.

维格堂东面全景
East View

维格堂历史照片
Vintage Image

维格堂东面入口
East Entrance

维格堂东面底层窗户
East Window (First Floor)

维格堂东立面
East Elevation

维格堂东面中部
Midsection of East Facade

维格堂南面窗户
South Window

维格堂南立面
South Elevation

维格堂南面大门
South Entrance

子实堂 ZISHI HALL

建筑年代: 1930年
原名: 子实堂
原功能: 学生宿舍
现名: 子实堂
现功能: 苏州大学人文社科研究院、学报编辑部等
历史沿革: 维格堂正对着的建筑是1930年落成的子实堂,为纪念存养书院、博习书院创办人华人曹子实而命名。子实堂是当时的学生宿舍,楼高四层,风格简朴、明快。

Year of construction: 1930
Original name: Zishi Hall
Original usage: student dormitory
Current name: Zishi Hall
Current usage: School of Social Science, newsroom for the school magazine
History: Zishi Hall was built in 1930 and dedicated to the memory of Zishi Cao, who was the founder of both Cunyang Academy and Po-hsi Academy. The four-storey building, originally used as a dormitory is of a simple, yet lively style.

子实堂历史照片
Vintage Image

子实堂西面全景
West View

子实堂西立面
West Elevation

子实堂西面中部
Midsection of West Facade

子实堂西立面中部
Midsection of West Elevation

子实堂西面入口
West Entrance

博物馆 THE SOOCHOW UNIVERSITY MUSEUM

建筑年代:1934—1937年

原名:体育馆

原功能:游泳池、体育馆

现名:苏州大学博物馆

现功能:展览、陈列

历史沿革:东吴大学体育馆又名司马德体育馆,由游泳馆和体育馆两部分组成。这是为了纪念东吴体育创始人——美国人司马德。

这座带耳房的建筑平面呈“H”型,以清水红砖砌筑,其所采用的烧僵的砖块既节约成本又显得别有风味。这些大小不一、扭曲变形的深褐色砖块所构筑的壁柱与墙裙,砂浆砌缝宽大,凸显苍老厚重之感,粗糙的肌理产生与西方古典建筑中粗斫手法类似的效果,实属巧为利用、独具匠心的做法。建筑物内部是大空间结构,故采用了平缓的四坡歇山屋顶,主体部分屋架采用钢屋架和木斜梁支撑。窗洞高大,采用半圆券式样,保证了室内采光。建筑物砖墙瓦顶的基调,同样与周边建筑取得了协调。入口突出并饰以文艺复兴式线脚,上方有孔祥熙所题“体育馆”三字。

2010年5月,苏州大学建校110周年之际,基于司马德体育馆建成的苏州大学博物馆正式开馆。

Years of construction: 1934-1937
Original name: Gymnasium
Original usage: swimming pool, gymnasium
Current name: The Soochow University Museum
Current usage: historical exhibition
History: The “official” name of Dongwu University's Gymnasium was “The Smart Memorial Gymnasium” in memory of the founder of Dongwu Sports, Mr Smart from the United States. Over time it became known, simply, as “The Gymnasium”.

The plan of the building is in the shape of the letter “H”, and it once housed an indoor swimming pool and a gymnasium. The plain red brick which was used as the main construction material was economical and created a rather distinctive style. Brown-coloured bricks of various sizes and distorted shapes were used for the pilaster and wainscoting. The mortared seams were intentionally loose to create a rough, weathered effect similar to the craftsmanship of “Rough Chop” seen in classical western architecture. This demonstrated the architect's creativity in the use of the materials.

In order to create a large open space, the building utilized the gentle four-sloped hip-and-gable roof. The main structure is supported by a steel frame and a wood oblique beam. The window openings are large semicircular arches which allow ample natural light to penetrate the interior. The building fits well in the context of its surroundings with its brick wall and tile roof.

The entrance is enhanced by the Renaissance style architrave as well as the inscription “Gymnasium” in Chinese characters by Xiangxi Kong.

The building underwent extensive renovation and in May 2010, in its new guise as The Museum of Soochow University, opened to the public in celebration of the 110th Anniversary of the university.

原体育馆设计图
Design Sketch of Former Gymnasium

博物馆西面全景
West View

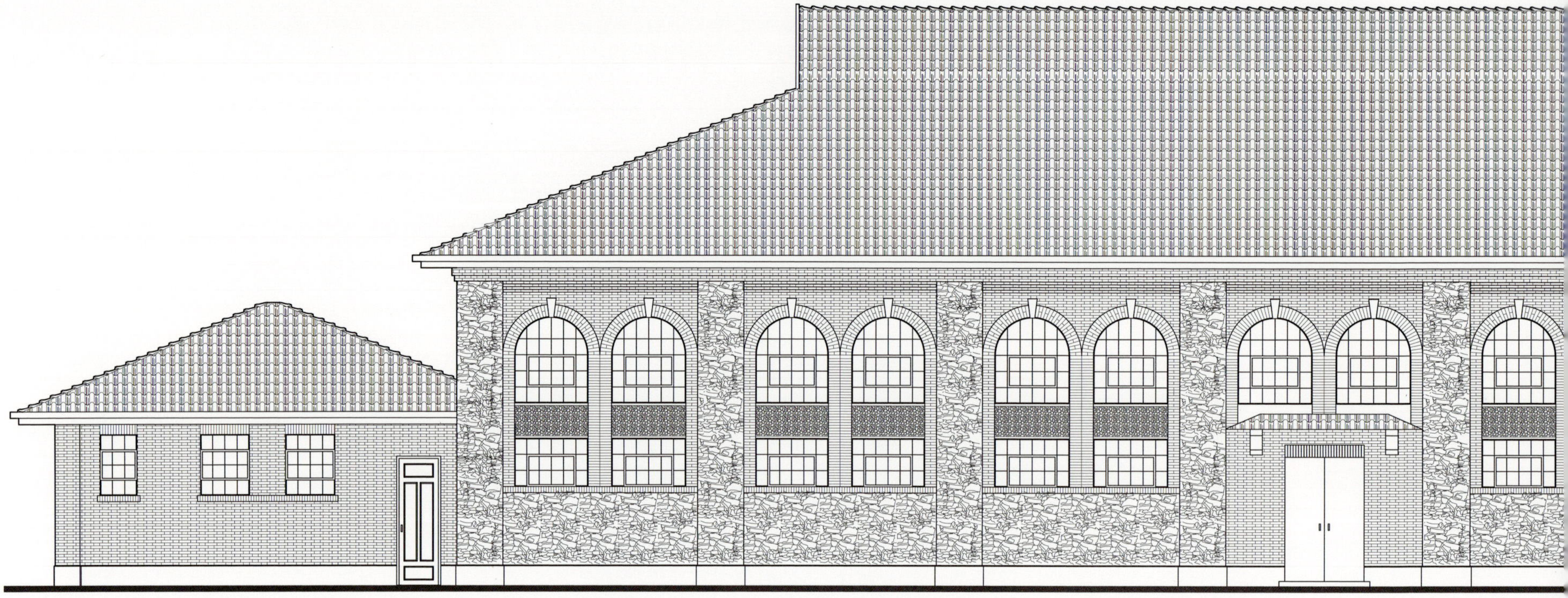

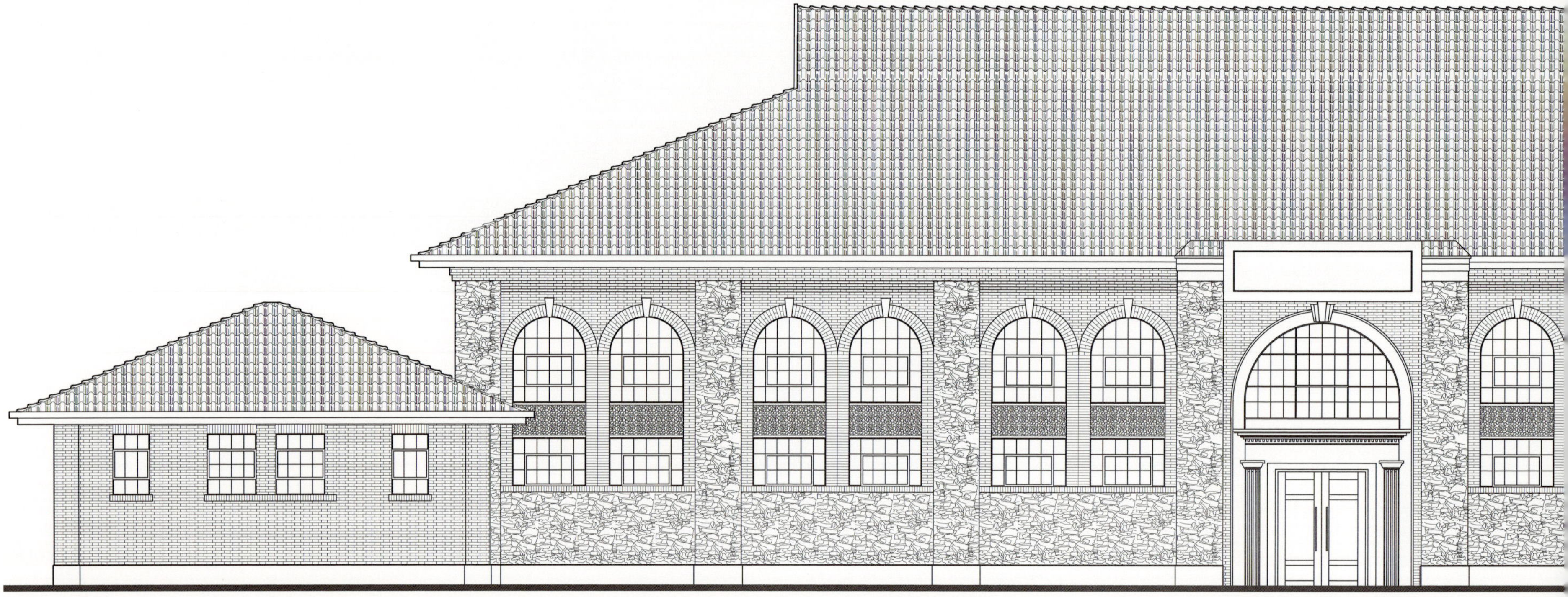

博物馆东立面
East Elevation

博物馆西立面
West Elevation

博物馆西面局部
Detail of West Facade

博物馆入口局部
Detail of Entrance

博物馆窗户局部
Detail of Window

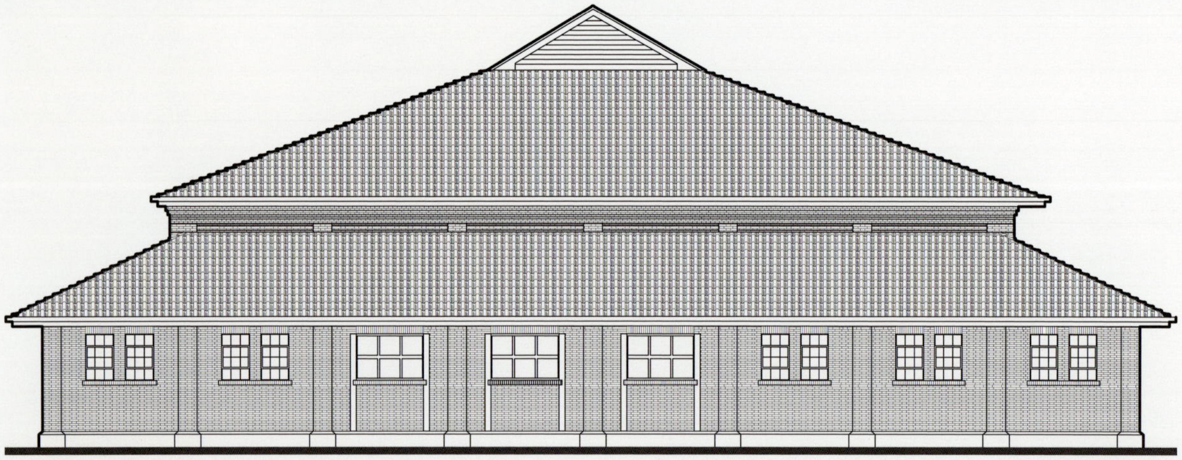

博物馆西立面
West Elevation

博物馆西面主入口
West Entrance

博物馆碎砖墙面
Rough Chop Brickwork

博物馆西立面主入口
West Entrance

博物馆西面窗户
West Window

春晖楼 CHUNHUI (SPRING) BUILDING

建筑年代:20世纪二三十年代
原功能:教师住宅
现名:春晖楼
现功能:外国语学院
历史沿革:与红楼会议中心相对的是四幢欧式风格的小别墅,原来是外教的住宿楼,建于上世纪二三十年代。新中国成立后其功能历经变迁,现名系2000年苏州大学百年校庆时征集所得,自东向西分别为春晖楼、秋韵楼、夏润楼、冬瑞楼。

Years of construction: 1920's-1930's
Original usage: faculty residence
Current name: Chunhui (Spring) Building
Current usage: School of Foreign Languages
History: On the opposite side of the Red Mansion Conference Centre stand four European style villas, which were built in the 1920's and 1930's and were once the residences for the foreign teachers. The function of the buildings changed after the founding of the People's Republic of China.
In 2000, during the celebration of the 100th Anniversary of the University, the villas were renamed after the seasons. From east to west: Spring Building, Autumn Building, Summer Building, and Winter Building.

春晖楼东南面全景
South-east View

春晖楼南面全景
South View

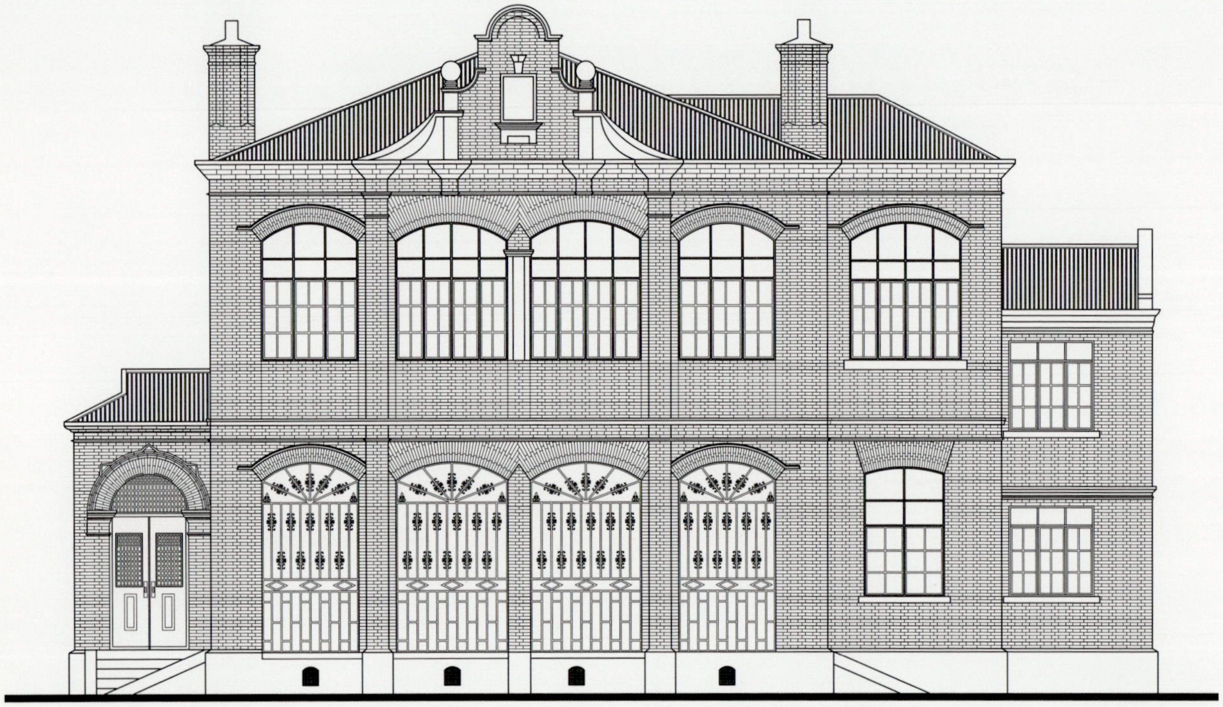

春晖楼南立面
South Elevation

春晖楼北面全景
North View

春晖楼西面局部
Detail of West Facade

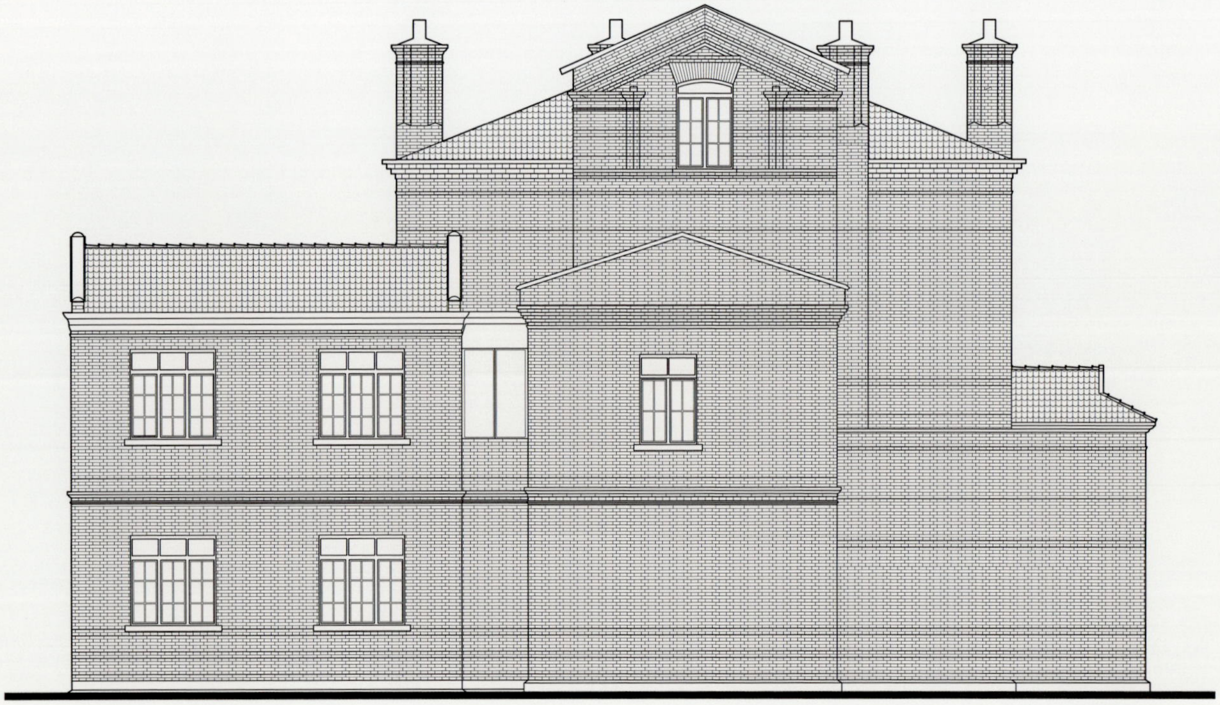

春晖楼北立面
North Elevation

春晖楼东面烟囱
East Chimney

春晖楼东面局部
Detail of East Facade

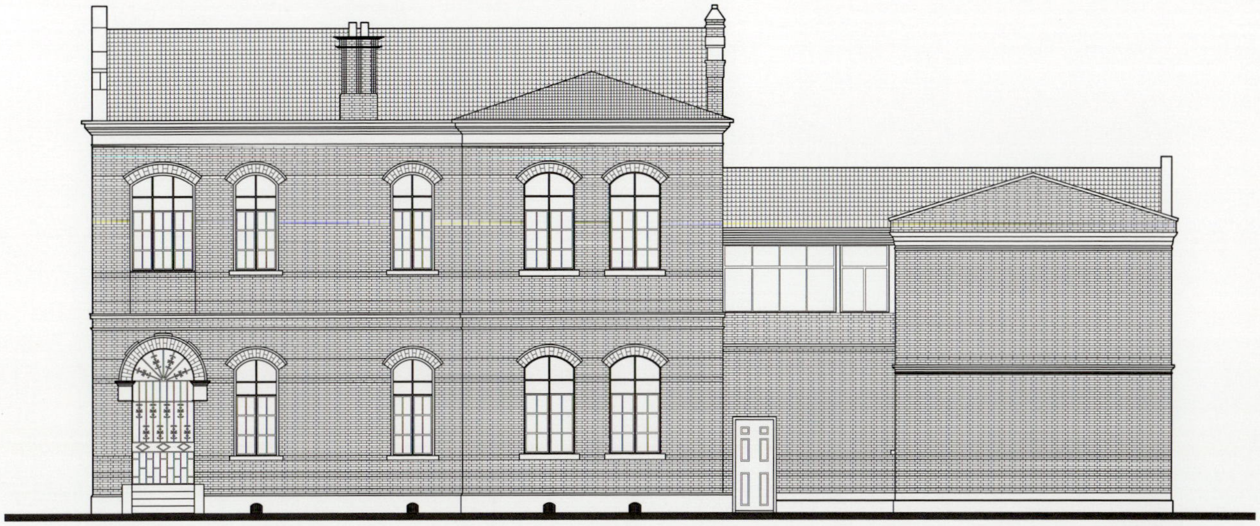

春晖楼东立面
East Elevation

春晖楼东面入口
East Entrance

春晖楼西立面
West Elevation

春晖楼南立面局部
Detail of South Elevation

春晖楼南面山花
South Pediment

春晖楼西南面全景
South-west View

春晖楼西面门斗
West Porch

夏润楼　XIARUN (SUMMER) BUILDING

建筑年代:20世纪二三十年代

原功能:教师住宅

现名:夏润楼

现功能:行政办公

历史沿革:与红楼会议中心相对的是四幢欧式风格的小别墅,原来是外教的住宿楼,建于上世纪二三十年代。新中国成立后其功能历经变迁,现名系2000年苏州大学百年校庆时征集所得,自东向西分别为春晖楼、秋韵楼、夏润楼、冬瑞楼。

Years of construction: 1920's-1930's

Original usage: faculty residence

Current name: Xiarun (Summer) Building

Current usage: administration offices

History: On the opposite side of the Red Mansion Conference Centre stand four European style villas, which were built in the 1920's and 1930's and were once the residences for the foreign teachers. The function of the buildings changed after the founding of the People's Republic of China.

In 2000, during the celebration of the 100th Anniversary of the University, the villas were renamed after the seasons. From east to west: Spring Building, Autumn Building, Summer Building, and Winter Building.

夏润楼北面全景
North View

夏润楼西南面全景
South-west View

夏润楼东立面
East Elevation

夏润楼西立面
West Elevation

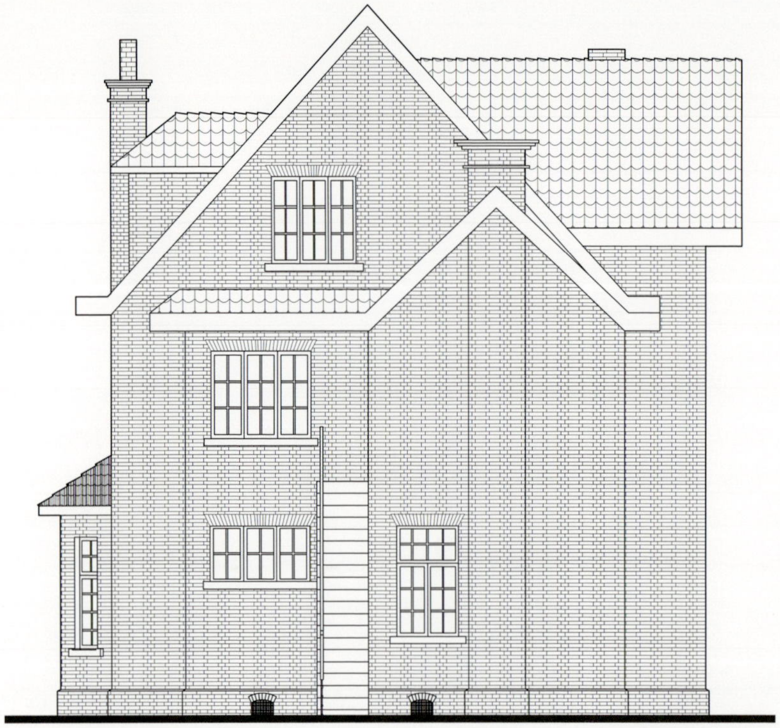

秋韵楼北立面
North Elevation

秋韵楼南面火焰券窗口
Ogee Arch of South Window

秋韵楼西立面
West Elevation

秋韵楼西面局部
Detail of West Facade

夏润楼西北面全景
North-west View

夏润楼南面窗户
South Window

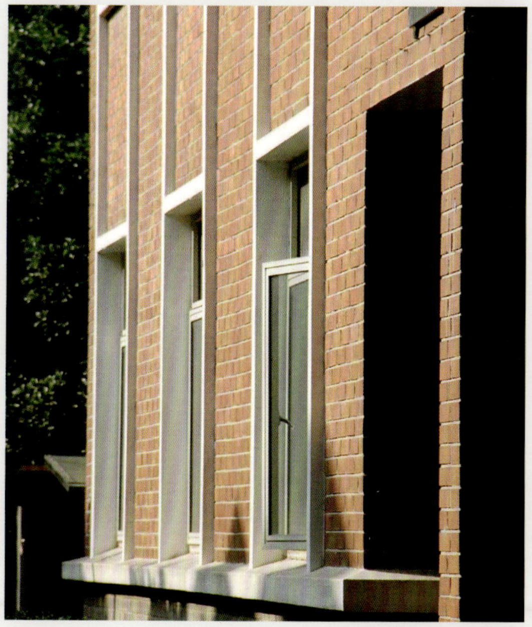

夏润楼西面局部
Detail of West Facade

夏润楼南立面
South Elevation

夏润楼北立面
North Elevation

秋韵楼 QIUYUN（AUTUMN）BUILDING

建筑年代：20世纪二三十年代

原功能：教师住宅

现名：秋韵楼

现功能：行政办公

历史沿革：与红楼会议中心相对的是四幢欧式风格的小别墅，原来是外教的住宿楼，建于上世纪二三十年代。新中国成立后其功能历经变迁，现名系2000年苏州大学百年校庆时征集所得，自东向西分别为春晖楼、秋韵楼、夏润楼、冬瑞楼。

Years of construction: 1920's-1930's

Original usage: faculty residence

Current names: Qiuyun (Autumn) Building

Current usage: administration offices

History: On the opposite side of the Red Mansion Conference Centre stand four European style villas, which were built in the 1920's and 1930's and were once the residences for the foreign teachers. The function of the buildings changed after the founding of the People's Republic of China.

In 2000, during the celebration of the 100th Anniversary of the University, the villas were renamed after the seasons. From east to west: Spring Building, Autumn Building, Summer Building, and Winter Building.

秋韵楼东南面全景
South-east View

秋韵楼西面局部
Detail of West Facade

秋韵楼南面局部
Detail of South Facade

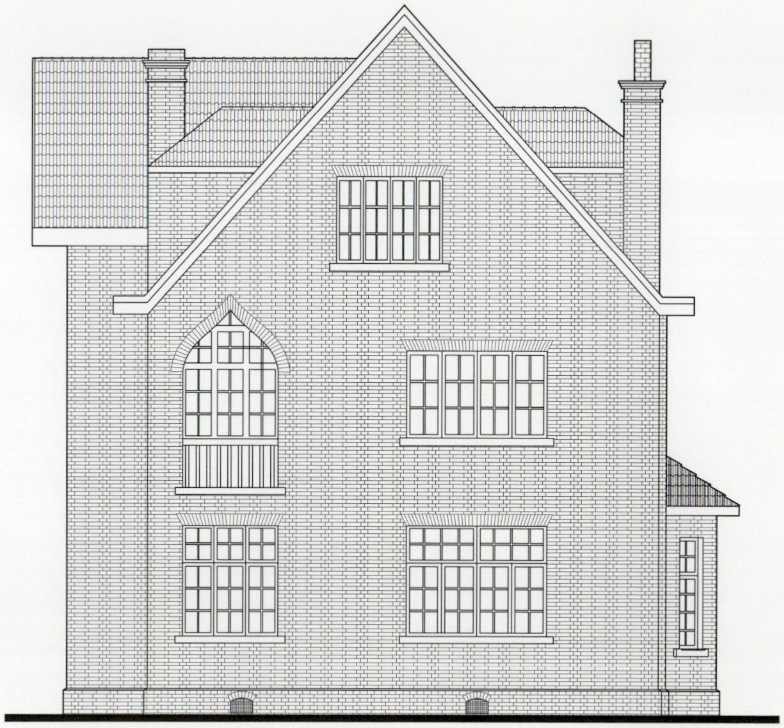

秋韵楼南立面
South Elevation

秋韵楼东面楼梯局部
Detail of East Stairs

秋韵楼东北面局部
Detail of North-east Facade

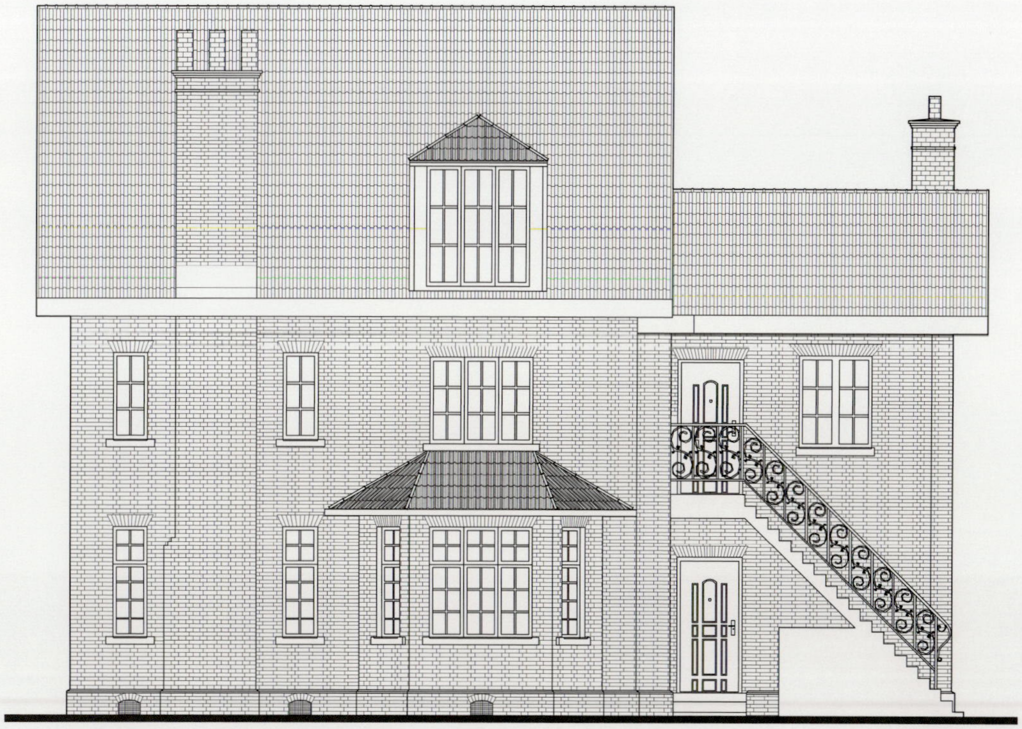

秋韵楼东立面
East Elevation

冬瑞楼 DONGRUI（WINTER）BUILDING

建筑年代：20世纪二三十年代
原功能：教师住宅
现名：冬瑞楼
现功能：东吴书画院
历史沿革：与红楼会议中心相对的是四幢欧式风格的小别墅，原来是外教的住宿楼，建于上世纪二三十年代。新中国成立后其功能历经变迁，现名系2000年苏州大学百年校庆时征集所得，自东向西分别为春晖楼、秋韵楼、夏润楼、冬瑞楼。

Years of construction: 1920's-1930's
Original usage: faculty residence
Current name: Dongrui (Winter) Building
Current usage: Dongwu Academy of Painting and Calligraphy
History: On the opposite side of the Red Mansion Conference Centre stand four European style villas, which were built in the 1920's and 1930's and were once the residences for the foreign teachers. The function of the buildings changed after the founding of the People's Republic of China.
In 2000, during the celebration of the 100th Anniversary of the University, the villas were renamed after the seasons. From east to west: Spring Building, Autumn Building, Summer Building, and Winter Building.

冬瑞楼西面山花局部
West Pediment

冬瑞楼西面全景
West View

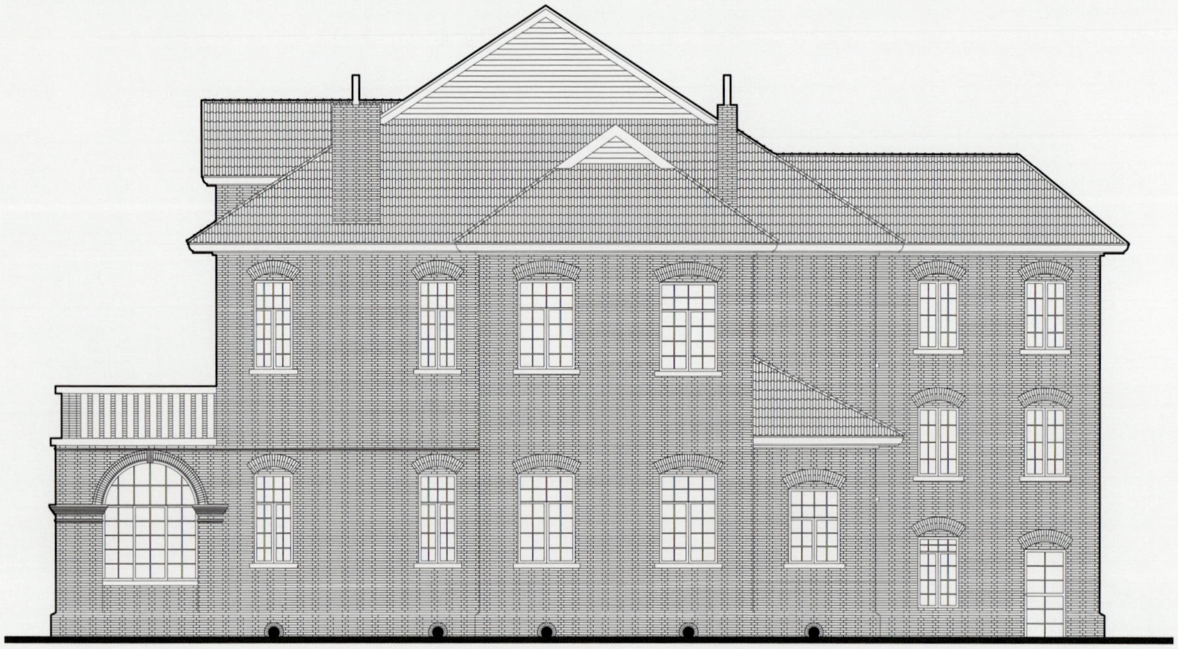

冬瑞楼东立面
East Elevation

冬瑞楼西立面
West Elevation

冬瑞楼西面局部
Detail of West Facade

冬瑞楼西面局部砖饰
Detail of West Facade Brick Decoration

冬瑞楼西面北侧局部
North-west Facade (Detail)

冬瑞楼西面局部
Detail of West Facade

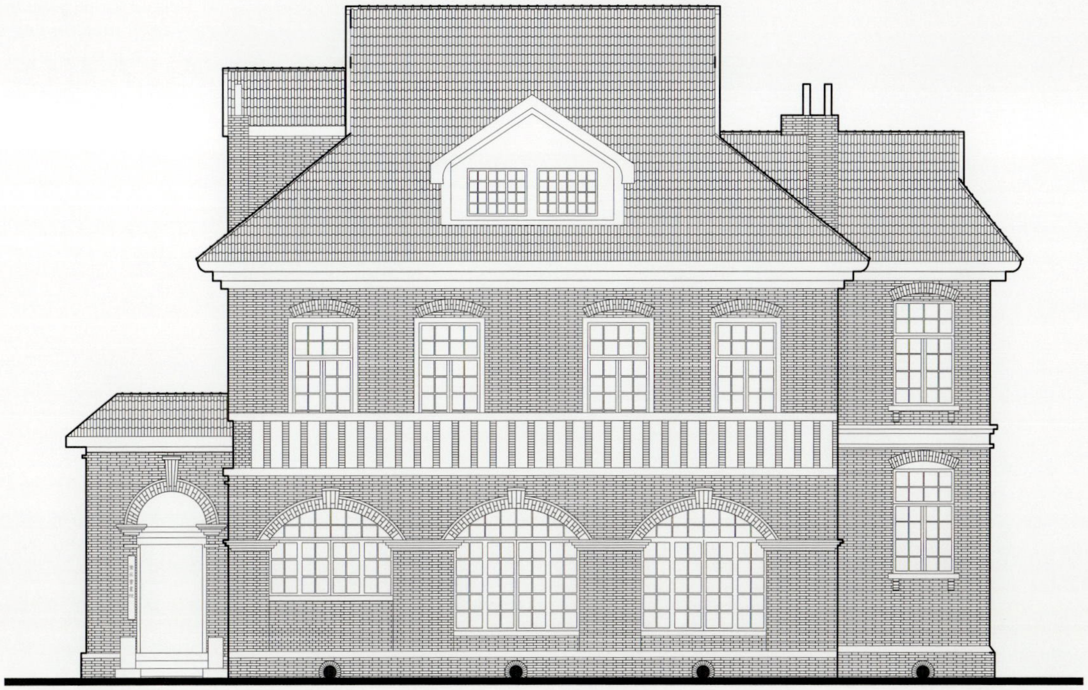

冬瑞楼南立面
South Elevation

冬瑞楼西面门斗
West Porch

冬瑞楼北立面
North Elevation

冬瑞楼西立面中部
Midsection of West Elevation

冬瑞楼西面中部
Midsection of West Facade

冬瑞楼西面花窗棂
West Tracery

崇远楼 CHONGYUAN BUILDING

设计年代：约1936年

建筑师：上海华盖建筑师事务所

原名：苏州景海女子师范学校

原功能：苏州景海女子师范学校校舍

现名：崇远楼

现功能：苏州大学外国语学院

历史沿革：崇远楼为砖混结构，民国早期建筑风格，它和红楼、敬贤堂一起作为苏州景海女子师范学校前身，是于光绪二十八年(1902年)在苏州天赐庄创办的景海女塾，原为纪念美国基督教监理公会女传教士海淑德(Laura Askew Haygood)而建。民国6年(1917年)改为苏州景海女子师范学校。新中国成立后，学校经分拆合并后不复存在。2004年12月23日，景海女师旧址被列为苏州市第五批公布的市级文物保护单位。

Year of design: around 1936

Designed by: Shanghai Huagai Architects

Original name: Jinghai Female Normal School

Original usage: dormitory

Current name: Chongyuan Building

Current usage: the School of Foreign Languages

History: Chongyuan Building is of early Mingguo architectural style, masonry structure. Together with the Red Mansion and Jingxian Hall, the Chongyuan Building was originally built to house the Jinghai Female Private School which was founded by Suzhou Tian Ci Zhuang in 1902. It commemorates Laura Askew Haygood, an American commissioner who was sent to Suzhou by the Women's Missionary Society of the Methodist Church. Later in 1917, the school was renamed, simply, "Jinghai Female Normal School". The school was split and re-united several times after the founding of the People's Republic of China, but now, it no longer exists. The old site of Jinghai Female Normal School was listed in the fifth batch of Municipal Heritage Sites in December 2004.

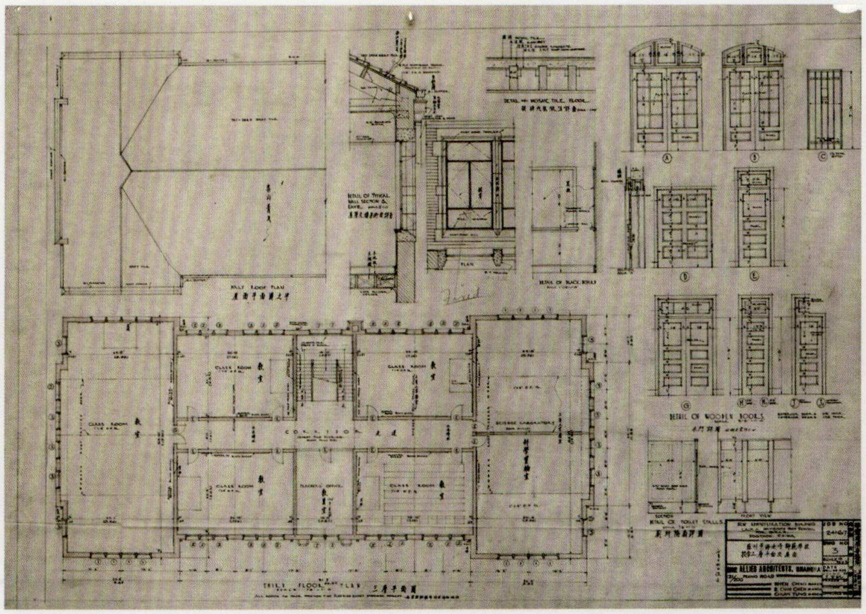

崇远楼平面图、大样图
Plan and Detail Drawings

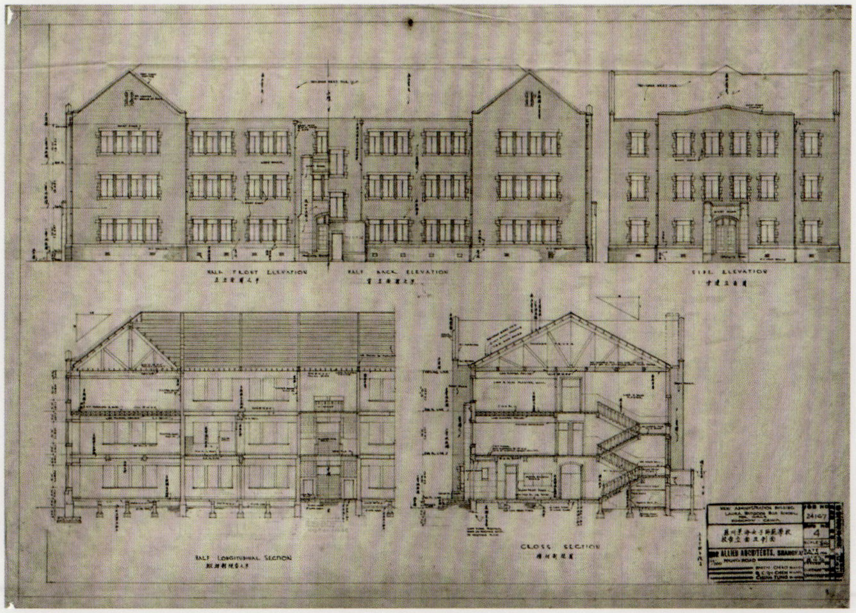

崇远楼立面图、剖面图
Elevation and Section

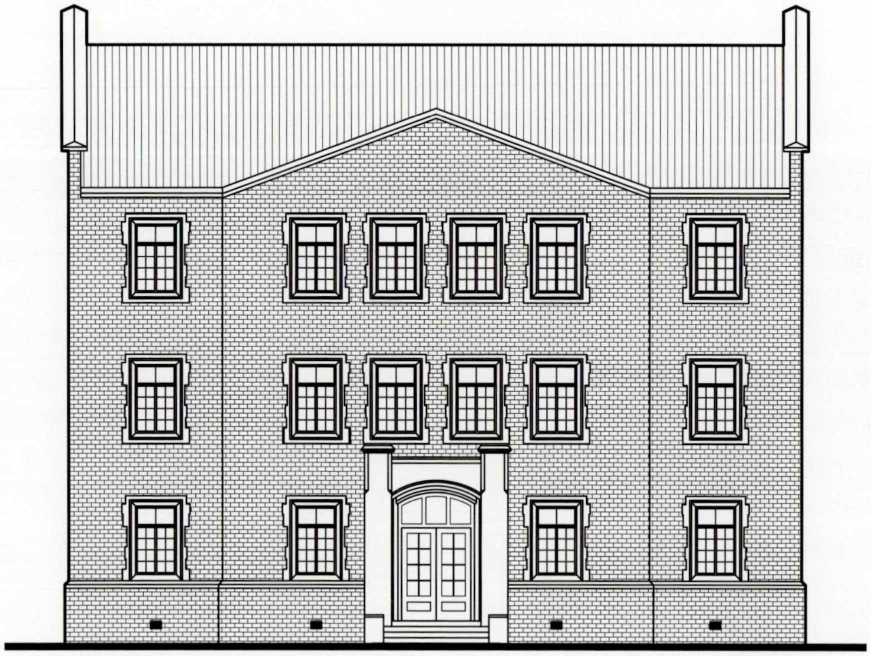

崇远楼东立面
East Elevation

崇远楼东面全景
East View

崇远楼东面入口局部
Detail of East Entrance

崇远楼南立面
South Elevation

崇远楼南面入口局部
Detail of South Entrance

敬贤堂　JINGXIAN HALL

设计年代：约1936年

建筑师：上海华盖建筑师事务所

原名：苏州景海女子师范学校

原功能：苏州景海女子师范学校礼堂

现名：敬贤堂

现功能：礼堂

历史沿革：敬贤堂为民国早期建筑风格，它原为景海女塾的礼堂并一直沿用至今。敬贤堂和红楼及崇远楼一起，作为苏州景海女子师范学校前身，是于光绪二十八年(1902年)在苏州天赐庄创办的景海女塾，原为纪念美国基督教监理公会女传教士海淑德(Laura Askew Haygood)而建。民国六年(1917年)改为景海女子师范学校。新中国成立后，学校经分拆合并后不复存在。2004年12月23日，景海女师旧址被列为苏州市第五批公布的市级文物保护单位。

Year of design: around 1936

Designed by: Shanghai Huagai Architects

Original name: Jinghai Female Normal School

Original usage: auditorium

Current name: Jingxian Hall

Current usage: auditorium

History: Jingxian Hall, which is of Mingguo architectural style, was originally built as an Auditorium for the Jinghai Female Normal School and retains its original function to this day.

Together with the Red Mansion and the Chongyuan Building, Jingxian Hall was built to house the Jinghai Female Private School which was founded by Suzhou Tian Ci Zhuang in 1902. It commemorates Laura Askew Haygood, an American commissioner who was sent to Suzhou by the Women's Missionary Society of the Methodist Church. Later in 1917, the school was renamed, simply, "Jinghai Female Normal School".

The school was split and re-united several times after the founding of the People's Republic of China, but now, it no longer exists. The old site of Jinghai Female Normal School was listed in the fifth batch of Municipal Heritage Sites in December 2004.

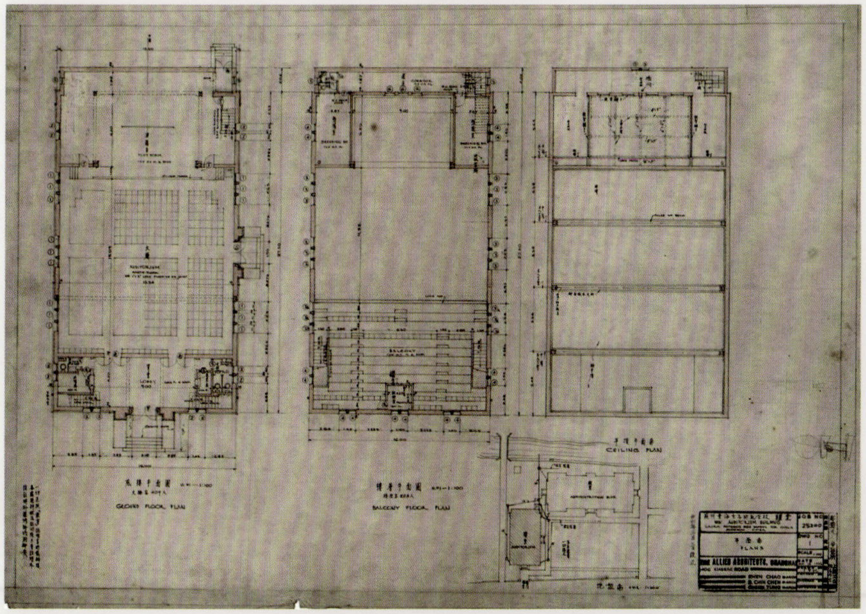

敬贤堂平面图
Plan Drawings

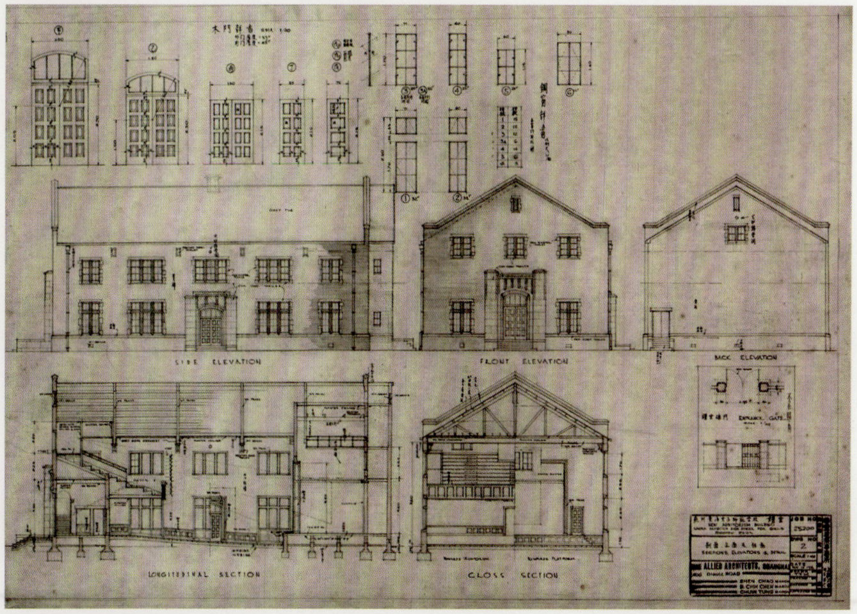

敬贤堂立面图、剖面图
Elevation and Section

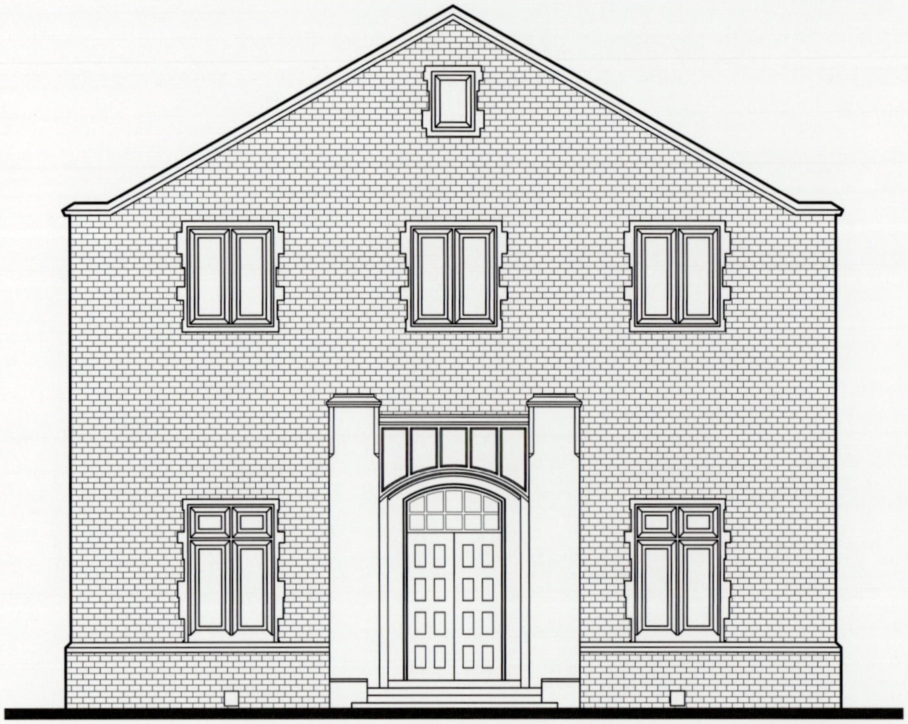

敬贤堂南立面
South Elevation

敬贤堂南面全景
South View

敬贤堂东面局部
Detail of East Facade

敬贤堂东立面
East Elevation

敬贤堂东面入口局部
Detail of East Entrance

文星阁 WENXING TOWER

建筑年代:明万历十七年至二十五年建，万历四十年重建

原名:文星阁

原功能:钟塔

又名:钟楼、方塔

现功能:苏州大学教工之家

历史沿革:始建于明万历十七年(1589年)至二十五年，附属于县学。万历四十年重建，稍向南移，即今所在。清代葑门彭氏曾于康熙二年(1663年)、乾隆十七年(1752年)出资重修，并于阁旁建桂香殿、朝元阁、时习堂、三贤堂等，成为名儒文士讲学会文的地方。

咸丰十年(1860年)至同治二年(1863年)，文星阁曾被太平军用作探望军情的望楼，附属殿阁斋堂毁于兵火。同治九年修阁，现阁南仅存的桂香殿为同治十一年修建。

阁楼分四层。下承三级青石方台基，高6.5米，底边宽14.4米，逐级向上递减，南面有石级作八字形，可从两侧拾级而登。上覆四角攒尖顶，翼角起翘，葫芦结顶。通高约28米。阁身平面呈正方形，各层四面辟为拱门，层间无腰檐平座。底层边宽8.6米，自下而上逐层作不甚明显的收分，形成有层次而又挺直的轮廓线。阁南尚存清代同治年间修建的桂香殿。

1982年，方塔被列为苏州市文物保护单位，2002年被列入江苏省文物保护单位。

Years of construction: 1589-1597(Ming Dynasty), rebuilt in 1612

Original name: Wenxing Tower

Original usage: Bell tower

Other names: Bell Tower, Square Tower

Current usage: faculty center

History: The Wenxing Tower was built in the 16th year of the reign of Emperor Wanli in the Ming Dynasty (1589). The current location was shifted a little to the south of the original tower when it was rebuilt in the 40th year of Wanli in the Ming Dynasty (1612).

In the 2nd year of Emperor Kangxi (1663) and the 17th year of Emperor Qianlong (1752) in the Qing Dynasty, the tower went through rehabilitations funded by the Peng family. At the same time, four affiliated buildings, the Guixiang Palace, the Chaoyuan Tower, Shixi Hall, and Sanxian Hall, were also built around the Tower. The complex was regarded as a popular gathering place for famous scholars.

During the 10th year of Emperor Xianfeng (1860) and the 2nd year of Emperor Tongzhi (1863), the Wenxing Tower was used as a military watchtower during the Taiping Rebellion (1850-1864). The affiliated buildings were destroyed during the Rebellion.

The Tower was renovated once again in the 9th year of Emperor Tongzhi (1870). The Guixianq Palace (to the south of the Tower) was rebuilt in the 11th year of Emperor Tongzhi (1872), and is the only other building in the complex to survive to this day.

The square four-storey Wenxing Tower is 28 meters tall, topped with a four-corner spire. It is divided into three levels with a 6.5 meter-high base made of blue stone. The base is 14.4 meters across, decreasing as the level rises. Access to the tower is from two directions with stairs on the south side.

On each floor, there is one arched door on each of the four sides. There is no waist eave flat between the floors. The length of the ground floor is 8.6 meters and each floor has an unnoticeable setback to achieve a layered and straight contour.

Wenxing Tower was listed as an historical and cultural site under the protection of the Soochow Government in 1982 and protection was raised in 2002 to that of the Jiangsu Province Government level.

文星阁西南面桂香殿
Guixiang Temple

文星阁东南面远景
South-east View

文星阁南面全景
South View

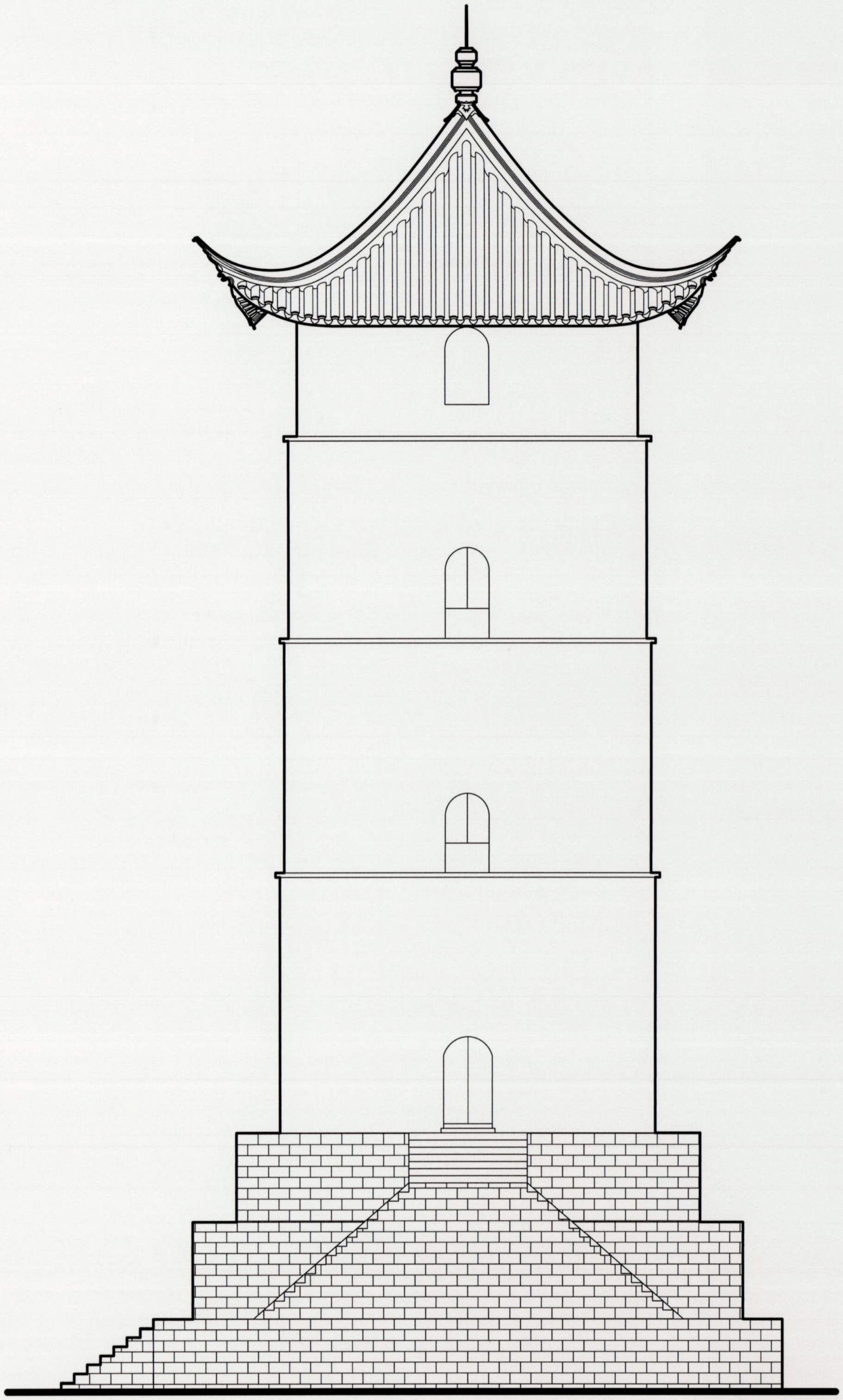

文星阁南立面
South Elevation

文星阁院墙门斗脊部
Ridge

文星阁院墙漏窗
Tracery

文星阁南面局部
Detail of South Facade

博爱楼 BO'AI BUILDING

建筑年代：1922年

建筑师：G.F.Ashley Architect

原名：博习医院

原功能：西医医院

现名：博爱楼

现功能：宿舍楼

历史沿革：博习医院是美国南方监理会于1883年在中国苏州创办的一所西医医院。博爱楼中段主体部分原为两层歇山顶的门诊大楼，始建于1922年，建造时使用了大量的"陆慕金砖"，即为皇宫烧制的细料方砖，并刻有相关年号以及苏州府督造字样。1936年的火灾烧去中式屋顶，当年依据西式病房大楼建筑风格将门诊楼重新翻建成为三层平顶西式房屋，平顶四周砌半镂空的女儿墙作围护。1954年，因门诊需要，在建筑东西两翼各添建两层建筑，形成目前的建筑格局。建筑造型方整，清水砖墙、欧式线脚，显现着民国时代建筑特有的中西合璧的风格。2004年，该建筑被列为苏州市文物保护单位。

新中国成立后，该建筑历经苏州市第一人民医院、苏州医学院附属医院、苏州卫生学校的变迁，目前为苏大附一院职工家属宿舍。随着建筑所在天赐庄历史街区的改造更新，未来这里将引入新的功能，为昔日历史建筑注入新的活力。

Year of construction: 1922

Architect: G.F. Ashley Architect

Original name: Soochow Hospital

Original usage: hospital of Western medicine

Current name: Bo'ai Building

Current usage: dormitory

History: Soochow Hospital was established in 1883 by the America South Methodist Church as a hospital of western medicine in Suzhou.

The central part of the original Bo'ai Building, built in 1922, was a two-storey outpatient service building with a hip-and-gable roof. The principle construction material, "Lumu golden Brick", was the same fine brick used in building the Imperial Palace and was carved with the year of manufacture and the characters for "Made by the Soochow Government".

A fire in 1936 destroyed the Chinese style roof. A three-storey new building was designed with a flat roof with a half-height parapet wall as the fencing, a common style for western style hospitals of that period.

In 1954, two additional two-storey structures were added on the east and the west sides of the main building, giving it its current appearance.

The building is regular and balanced with plain brick walls and a European style architrave to create a fusion of Chinese and Western styles unique in that period.

In 2004, the building was listed as an historical and cultural site under the protection of the Soochow Government.

Since the founding of the People's Republic of China, the ownership of the building has shifted with the time. Currently it is the staff apartment building of the First Affiliated Hospital of Soochow University. But as restorations continue within the Tian Ci Zhuang Historical District, where the Bo'ai Building is located, the building will be given a new life and a new function.

博爱楼历史照片
Vintage Image

博爱楼南面中段全景
Midsection of South Facade

博爱楼北面中段全景
Midsection of North Facade

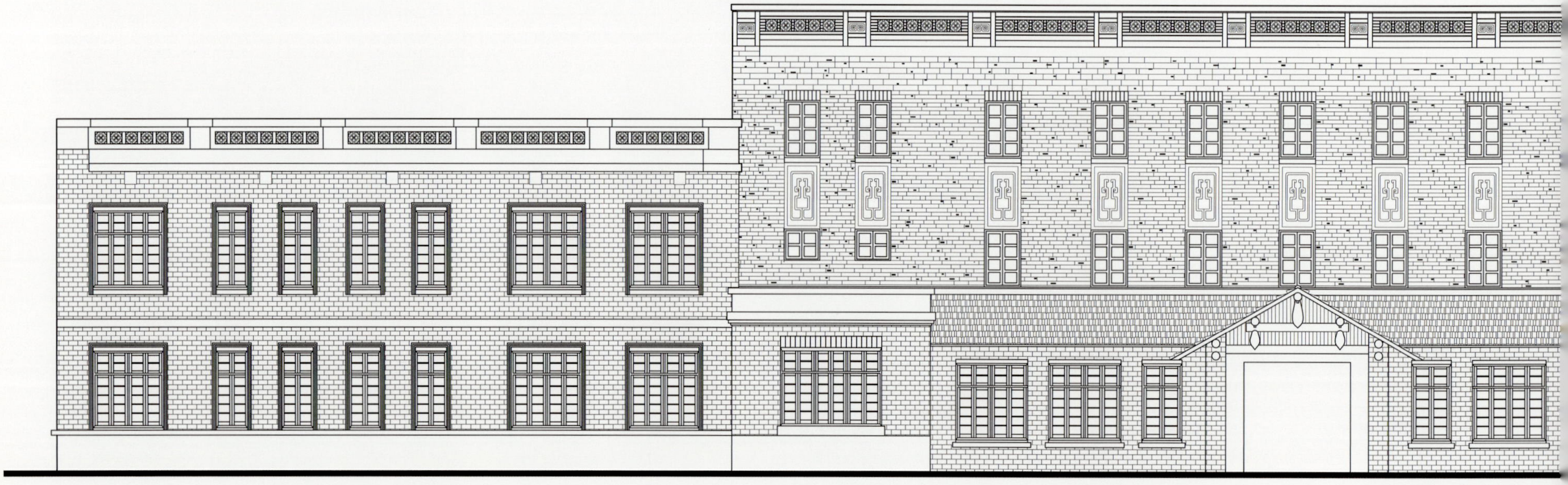

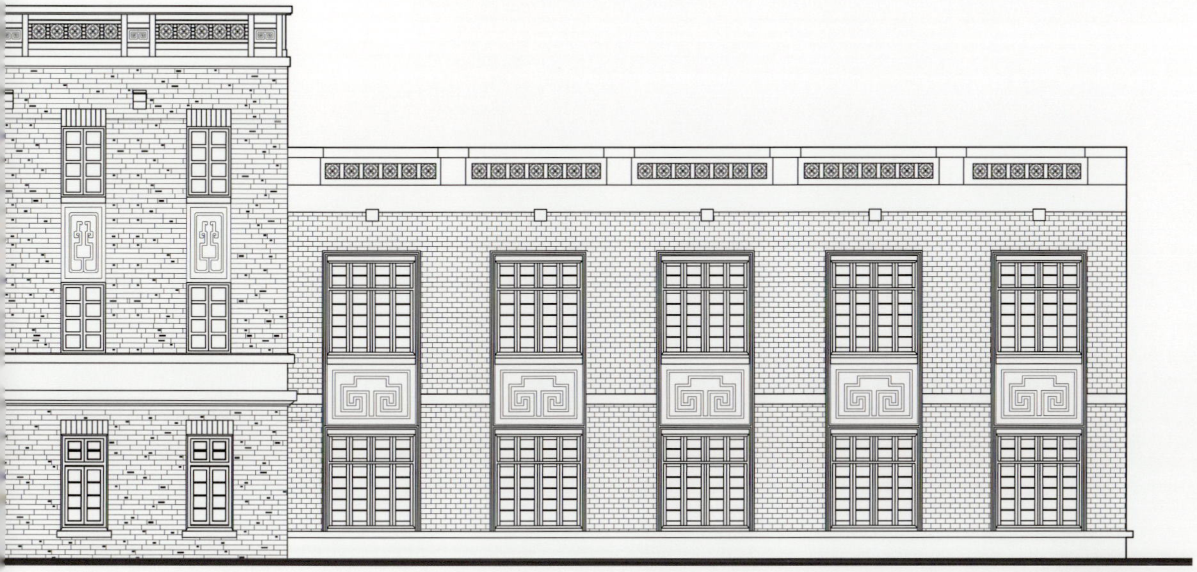

博爱楼南立面
South Elevation

博爱楼立面装饰纹样
Elevation Decorative Pattern

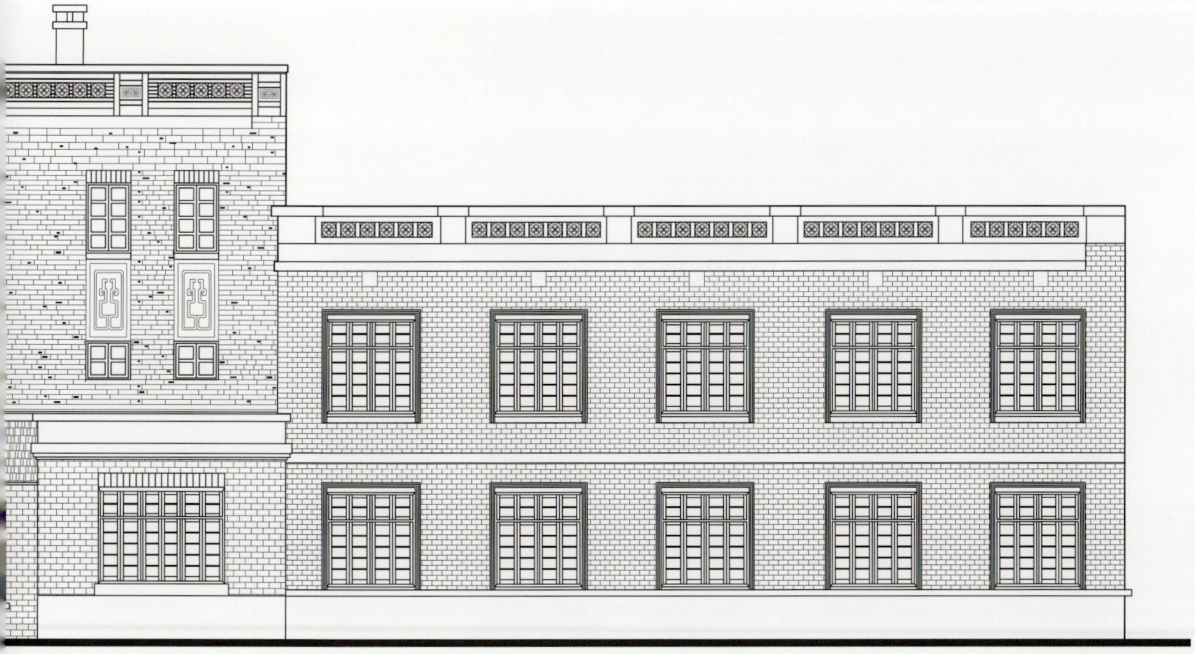

博爱楼北立面
North Elevation

博爱楼南面中段局部
Midsection of South Facade

博爱楼金砖铭文
Brick Inscriptions

博爱楼东立面
East Elevation

博爱楼南面局部
Detail of South Facade

博爱楼南面牛腿及檐口局部
Corbels and Eaves

博爱楼南面牛腿及檐口局部
Corbels and Eaves

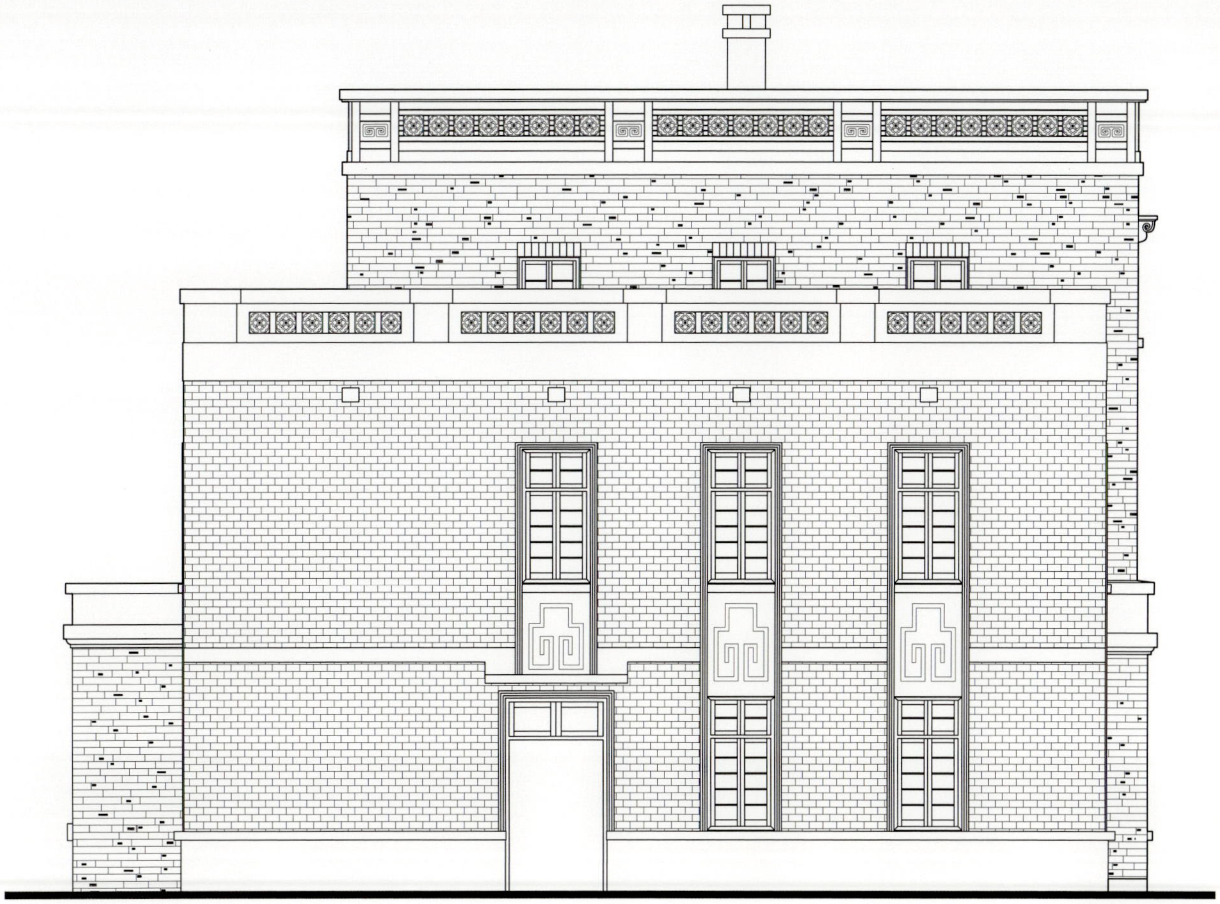

博爱楼西立面
West Elevation

圣约翰堂　ST. JOHN'S CHURCH

建筑年代：始建于1881年，1915年重建
建筑师：约翰·M.慕尔(美)
原名：首堂
原功能：教堂
现名：圣约翰堂
现功能：教堂
历史沿革：圣约翰堂前身为1881年美国监理会宣教士蓝柏、潘慎文所建，名为首堂，可容400人聚会。20世纪初，苏州的基督徒人数增加，到1915年，监理会拆除了首堂，新建了一座建筑面积1855平方米，有800个座位的西式教堂，并为纪念卫理公会的创始人约翰·卫斯理而改名为圣约翰堂。圣约翰堂成为当时苏州十大建筑之一。

1996年10月，圣约翰堂开始修复教堂工程，1998年5月竣工。2005年11月，圣约翰堂正式对外开放。

诺贝尔物理学奖得主李政道博士的祖父李仲覃牧师于1910年起任该堂首任华人主任牧师，李政道博士曾多次访问圣约翰堂。

Years of construction: 1881, rebuilt in 1915
Architect: John. M. Moore (U.S.)
Original name: Shou Tang
Original usage: church
Current name: St. John's Church
Current usage: church
History: Originally named Shou Tang, St. John's Church was built in 1881 by Bo Lan and Shenwen Pan who were missionaries of the American Methodist Church.

The church was large enough to accommodate 400 worshipers at one time. As the number of parishioners grew in the early 20th century, the Methodist Church decided to construct a new and larger (1,855 m^2) western style church to replace the old building. The new church was named St. John's to commemorate the founder of the Methodist Church, John Wesley.

The new church, which could accommodate 800 followers, was listed in the Top Ten Buildings of Soochow.

Between October 1996 and May 1998, St. John's Church underwent a major renovation, and reopened to the public in November 2005.

The Nobel Prize Winner for Physics Tsung Dao Lee's grandfather was the first Chinese priest of the church in 1910. Dr. Lee has visited the church several times.

圣约翰堂西南面全景
South-west View

圣约翰堂西面全景
West View

圣约翰堂南面全景
South View

圣约翰堂南立面
South Elevation

圣约翰堂南面塔楼局部
Detail of South Tower

圣约翰堂南面窗棂
South Window

圣约翰堂南立面局部
Detail of South Elevation

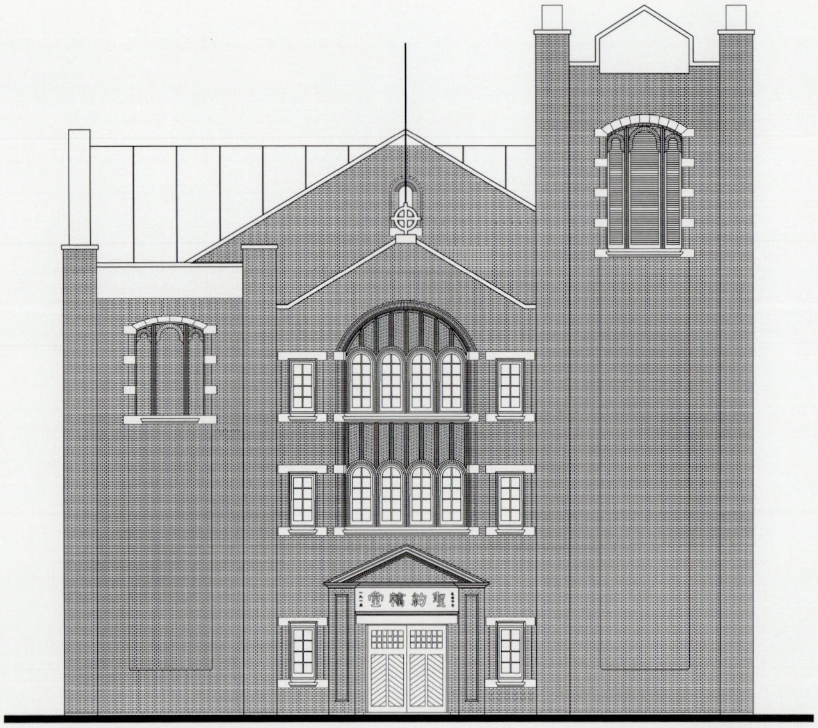

圣约翰堂西立面
West Elevation

圣约翰堂北立面
North Elevation

圣约翰堂西面盲窗
West Blind Tracery

圣约翰堂西面花窗棂
West Tracery

圣约翰堂西面门廊
West Porch

圣约翰堂西面山花局部
Detail of West Pediment

圣约翰堂西面中部
Midsection of West Facade

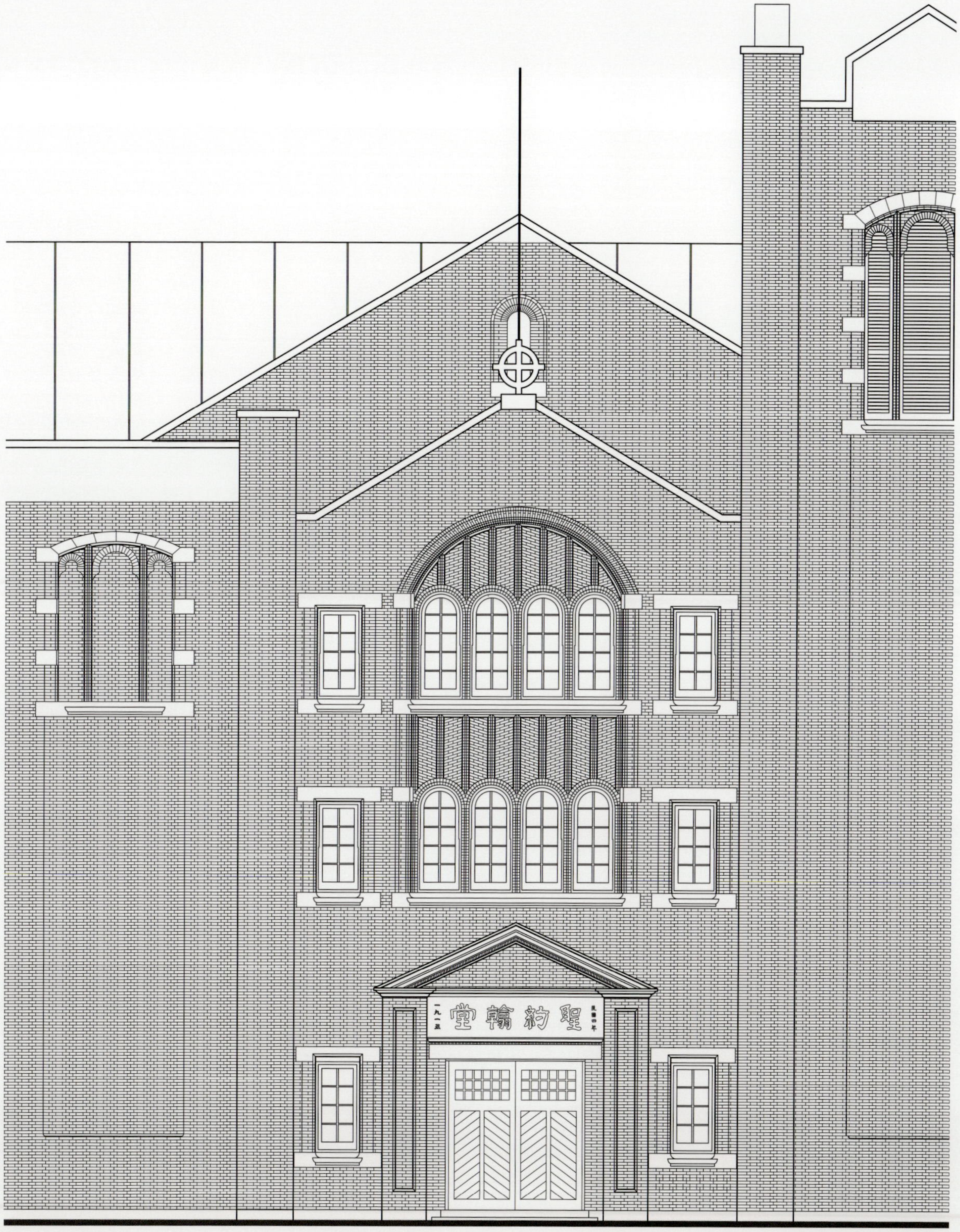

圣约翰堂西立面中部
Midsection of West Elevation

可园　KE GARDEN

建筑年代:初建于五代

原名:韩园(南宋)、近山林(清)、乐园(清)等

原功能:书院、图书馆

现名:可园

现功能:园林

历史沿革:可园基址五代时属吴越中吴军节度使孙承祐池馆,北宋时系沧浪亭的一部分,南宋时为韩园宅邸,元、明时其地并入大云庵。

后历经变迁,至清道光七年(1827年),布政使、江苏巡抚梁章钜重加修葺,划归正谊书院,易名"可园"。

1914年,设江苏省立苏州图书馆。可园曾经是江苏省立医学专门学校(1912年)、苏南工业专科学校(1951年)、苏南航空专科学校(1956年)、苏州医学院(1957年)校园一隅。现址位于苏州大学南校区。

可园现有面积5700平方米。水池居中,建筑疏朗,曲槛回廊,水木明瑟,庭宇清旷。建筑多为清光绪十四年(1888年)至民国时所构。主厅杏林堂在池北,三间歇山式四面厅,四周环廊。堂北土阜一丘,上立浩歌亭,为四角攒尖方亭。池西北有藏书楼博约楼,面阔五间。园中种植梅树逾百株。

Years of construction: originally built in the Five Dynasties period (907-960)

Historical names: Han Garden (South Song Dynasty), Jinshanlin Garden (Qing Dynasty), Happiness Garden (Qing Dynasty)

Original usage: academy of classical learning, library

Current name: Ke Garden

Current usage: decorative garden

History: Ke Garden has a long history. In the Five Dynasties period (907-960), the site of Ke Garden belonged to Chengyou Sun, an official of the State of Wu. During the Northern Song Dynasty (960-1127), it became part of the Canglang Pavilion. The Southern Song Dynasty (1127-1279) saw the property transferred to the Han family. And in the Yuan and Ming Dynasties, it was acquired by the Dayun Nunnery.

In 1827 the garden underwent a major renovation by Zhangju Liang, an official of Jiangsu Province and it was given its current name, Ke Garden, when it became part of Zhengyi Academy of Classical Learning.

In 1914, it was transformed into the Soochow Library. In the following years, as its surroundings changed, Ke Garden was attached to a variety of schools and academies: The Jiangsu School of Medicine (1912), The Sunan Academy of Technology (1951), The Sunan Aviation Academy (1956), The Soochow School of Medicine (1957). Ke Garden now lies within the South Campus of Soochow University.

The buildings, corridors, pavilions and plants that make up the 5,700 m^2 of Ke Garden are laid out around a small lake in a balanced way.

The buildings were built between the Qing Dynasty and the Republic of China period. The main hall, Xinglin Tang, is located on the north side of the lake. On the north side of the hall is a hill, on the top of which stands Haoge Pavilion, a four-cornered spire. To the north-west of the lake is the Boyue Library Building with five spaces between columns. There are more than one hundred Chinese plum trees in Ke Garden.

可园挹清堂
Yiqing Hall

可园小西湖月门洞
Moon Gate

可园濯缨处
Zhuoying House

可园花窗
Leaking Window

可园土丘及浩歌亭
Haoge Pavilion

可园入口砖雕门楼
Brick Carving Gateway

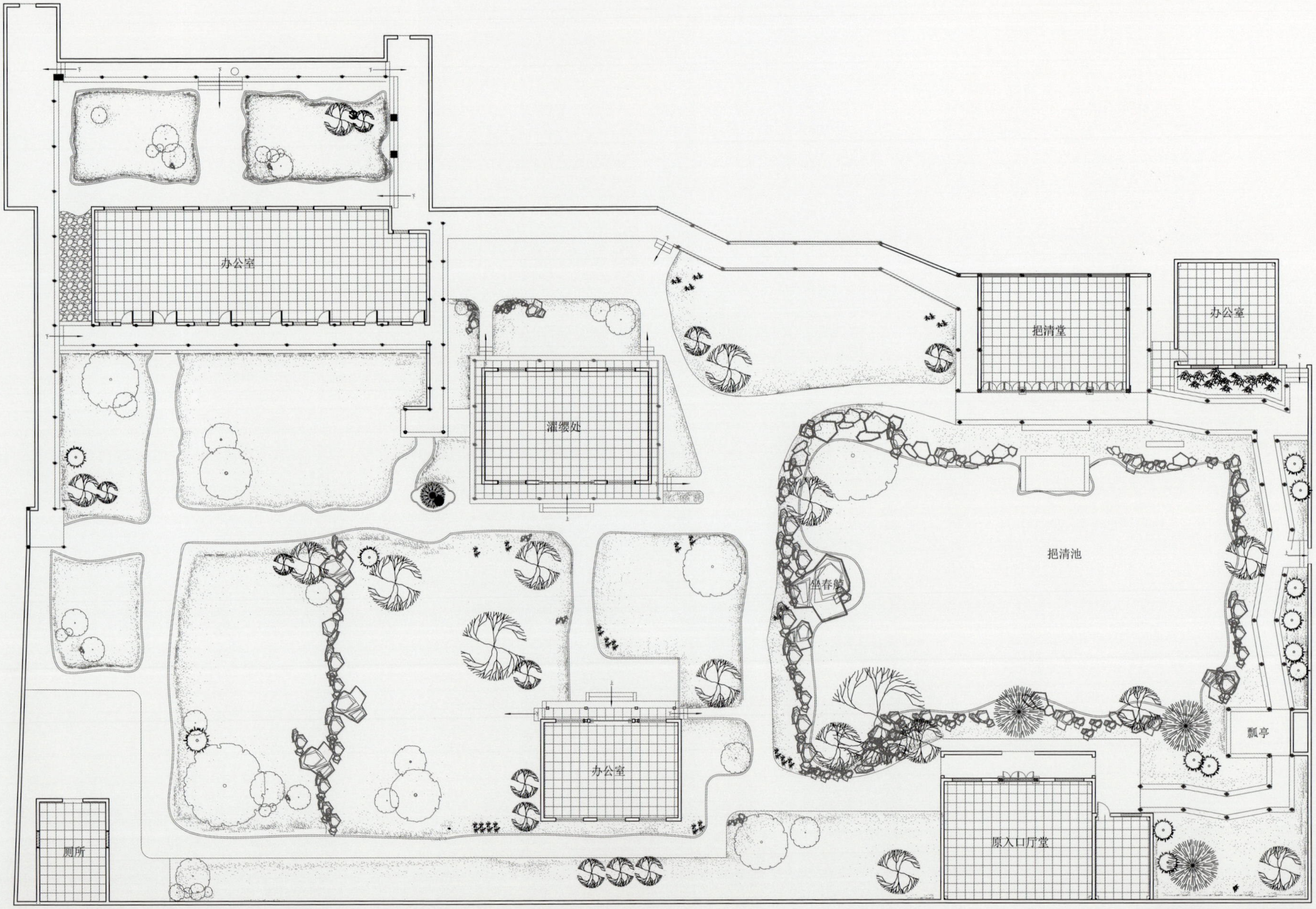

可园核心区域平面图
Site Plan

可园挹清池
Yiqing Pool

可园挹清堂及连廊
Colonnade of Yiqing Hall

可园一隅
A Corner

可园浩歌亭匾额
Horizontal Board of Haoge Pavilion

可园挹清堂飞檐转角
Upturned Eaves of Yiqing Hall

可园挹清堂山花
Pediment of Yiqing Hall

致 谢

本书的出版感谢苏州大学金螳螂建筑与城市环境学院2010级建筑学专业、2010级室内设计专业全体同学所做的建筑测绘及史料整理工作。

感谢苏州大学档案馆钱万里馆长、袁春荣老师为相关校史资料及历史图片的搜集整理提供的热情帮助。

感谢苏州大学金螳螂建筑与城市环境学院滕育梅教授所做的文字翻译工作，感谢苏州大学外语学院外籍教师 David Kindred 所做的文字校对工作。

感谢苏州大学金螳螂建筑与城市环境学院陈书尧、杨晓轩、陆原上骏、顾天玲同学所做的建筑摄影工作。

感谢苏州大学校史研究室王国平教授、苏州大学附属第一医院王馨荣教授、苏州市园林设计院沈贤成高级工程师为相关建筑史料的搜集和考证提供的热情帮助。

感谢苏州大学原党委办公室王洪法主任、原校长办公室于毓兰副主任、党委宣传部孙宁华副部长、原发展办公室薛辉主任、金螳螂建筑与城市环境学院查佐明书记为本书的出版提供的宝贵意见和协助。

ACKNOWLEDGMENTS

We wish to extend our thanks to all the students of 2010 Architecture and Interior Design from the School of Architecture and Urban Planning at Soochow University. Their hard work on surveying and historical data collection is greatly appreciated.

We thank Wanli Qian and Chunrong Yuan, the Librarians of Soochow University, for their enthusiastic support on the collection of vintage pictures and historical information.

Special thanks to Yumei Teng, professor of the School of Architecture and Urban Planning, for her hard work on English translation; and to David Kindred, English teacher of the School of Foreign Languages at Soochow University, for his dedication on English editing.

Thanks to Shurao Chen, Xiaoxuan Yang, Shangjun Luyuan, and Tianling Gu, the students of the School of Architecture and Urban Planning, for their photography.

Thanks also to professor Guoping Wang, Xinrong Wang at Soochow University, and Xiancheng Shen, senior Engineer at the Suzhou Institute of Landscape Architecture Design Co Ltd., Hongfa Wang, Yulan Yu, Ninghua Sun, Hui Xue, Zuoming Zha at Soochow University, for their enthusiastic support and precious advice.